TECHNOLOGY AND APPLICATION OF MICROBIAL FUEL CELLS

Edited by **Chin-Tsan Wang**

Technology and Application of Microbial Fuel Cells

http://dx.doi.org/10.5772/57200

Edited by Chin-Tsan Wang

Contributors

Chin-Tsan Wang, Oleksandr Bilyy, Oresta Vasyliv, Svitlana Hnatush, Rivka Cahan, Katalin Bélafi-Bakó, Aishwarya Mahadevan, Duminda Gunawardena, Sandun Fernando

Published by InTech

Janeza Trdine 9, 51000 Rijeka, Croatia

Publishing Process Manager Iva Lipovic

Technical Editor InTech DTP team

Cover InTech Design team

First published July, 2014

Printed in Croatia

Additional hard copies can be obtained from orders@intechopen.com

Technology and Application of Microbial Fuel Cells, Edited by Chin-Tsan Wang

p. cm.

ISBN 978-953-51-1627-1

For EU product safety concerns:
IN TECH d.o.o., Prolaz Marije Krucifikse Kozulić 3, 51000 Rijeka, Croatia,
info@intechopen.com or visit our website at intechopen.com.

Meet the editor

Professor Wang was born in 1970 and graduated from the Department of Aeronautics and Astronautics, National Cheng Kung University, Taiwan, with BS, MS and PhD qualifications undertaken between 1992 an 2000. He currently works at National I Lan University, Taiwan. As for his present areas of study, Microbial fuel cells, Micro fluidic devices and fuel cell flow slabs are the major fields. There are about 60 journal papers, 14 patents and 2 book chapter author contributions that can be attributed to his research performance to date. In addition, Professor Wang was awarded the honor of University Distinguished Professor in 2011 and presented with a National Academic Award in 2013.

Contents

Preface

It is my great honor to act as the editor of this book and be the author of the first chapter. This publication is made up of contributions from five experts in the field of Microbial fuel cells (MFCs), which have driven forth significant improvements in both the reliability and efficiency of MFCs and hence their future use.

It is known that microbial fuel cells were first created in 1911 to generate electrical power. In recent years, due to the environmental pollution and energy crises, the microbial fuel cell, with its recycling characteristics, has regained public attention due to its application to water treatment as well as being an asset for its ability to generate electrical power. However, MFCs are still confronted with the issue of having a low power rate generation and much effort has been devoted to the improvement of MFC performance. Now, this book will provide a platform for transdisciplinary conversation and understanding of the technical and business challenges of microbiology and green-energy research.

Distinguished Professor, Chin-Tsan Wang
Department of Mechanical and Electromechanical Engineering,
National I Lan University,
Taiwan

Chapter 1

Flow Control in Microbial Fuel Cells

Chin-Tsan Wang

Additional information is available at the end of the chapter

http://dx.doi.org/10.5772/58346

1. Introduction

1.1. Mass transfer of MFCs

In Microbial Fuel Cells (MFCs), a mass transfer impact might be larger than expected because it not only effects the mass transfer losses (substrate flux [1, 2] and removal of waste [2]), but also effects the activation losses (mediator [3-5]) and the ohmic losses (proton transfer [6]). However, the convection/diffusion of the reactant transport from the bulk (reactant transport), which can be enhanced by increasing the concentration of reactants and the mean flow rate [7], is able to enhance the power density of the MFCs by effective flow control. Now, a common flow control method of utilizing a magnetic stirrer bar/stirring plate was used to mix the solution of anolytes [8-10]. Nonetheless, some superior flow controls by applying a micro-mixer or micro-channel in the MFCs, which can enhance the power performance [11] in the cases of better solution mixing and without extra power supply, will be addressed in this chapter.

2. Using the magnetic stirrer bar

2.1. Different stirring speed strategies [9]

Different stirring speeds (100, 250, 380 rpm) in the anode of photo microbial fuel cells (PMFCs) have been studied by Lan et al., [9], Lan indicated that the internal resistance fell rapidly at 380 rpm and reached a maximum power density of 1.82mW/m^2. Using a high stirring speed (380 rpm) to mix the anolytes not only decreased the competition of microorganisms for delivering the electrons to the surface of the anode electrode, but also reduced the thickness of the mass transfer boundary layer (MTBL) [9]. However, an external power supply needed for the stirrer bar in PMFCs will weaken it at a larger scale of real application in the future.

3. Applying the micro-mixer or micro-channel in MFCs

The effect of the Reynolds number plays an important role in the flow control of the flow channel because it can affect the flow field of the MFC system. The Reynolds number, Re, shown in equation (1) is defined as the ratio of inertial forces to viscous forces [12]:

$$\mathrm{Re} = \frac{\rho V_0 D_0}{\mu} \tag{1}$$

Where ϱ is the density of fluid, V_0 means the average velocity of inlet flow channel, D_0 is the inlet width of flow channel and μ is the fluid dynamic viscosity coefficient.

In addition, the flow channel is designed to reduce the mass transfer losses of the MFCs. Generally speaking, a higher flow mixing is, the better the power performance of a MFC will be. Therefore, the mixing efficiency, $\varepsilon_{mixing,}$ defined in equation (2) will be used and necessary to identify whether a reactant is uniformly mixed with the anolyte [13]:

$$\varepsilon_{\mathrm{mixing}} = 1 - \frac{1}{L}\int_0^L \left| \frac{X_{\mathrm{Ax,outlet}} - 0.5}{X_{\mathrm{Amax}} - 0.5} \right| dx \tag{2}$$

Where ε_{mixing} is the mixing efficiency, X_{Amax} the maximum concentration of substance A, with the value always set as 1, $X_{Ax,outlet}$ the concentration of substance A at an outlet position, and L is the width of the outlet.

3.1. Convergent flow channel [14]

In this paragraph, a simple flow chamber with a higher flow mixing in MFCs will be addressed. The concept of a convergent flow channel with/without plates (as shown in Fig. 1) applies to the sediment microbial fuel cells (SMFCs) with three Re numbers (Re=30, 60, 90), which have been studied by Wang et al., with Fig. 2 showing the structure diagram of the SMFCs. Compared with a traditional MFC (without channel), the MFC with the convergent flow channel can produce two flow mechanisms (convection and vortex formed, as shown in Fig. 3 and Fig. 4) to enhance the mass transfer, cause the two flow mechanisms to make the substrate move tensely and the mixing more uniform. The electron produced and Hydrogen ion (H^+) will move faster and more actively to reach their objective, which can reduce the mass transfer losses and enhance the electricity production. On the other hand, the performance of a MFC with plates seems to be much more stable than in the case of without plates, causing the vortex flow to limit the moving scale of the substrate and make the transfer speed of the electrons and H^+more uniform, which can thus stabilize the power generation. However, the reason why the existing flow with plates lowers the open circuit voltage (OCV) and the power density has not been clarified clearly, but the other two vortexes (as shown in Fig. 3) caused by the flow plates may prevent some particles from escaping from the vortex scale and cause fewer electrons and H^+participating in the solution.

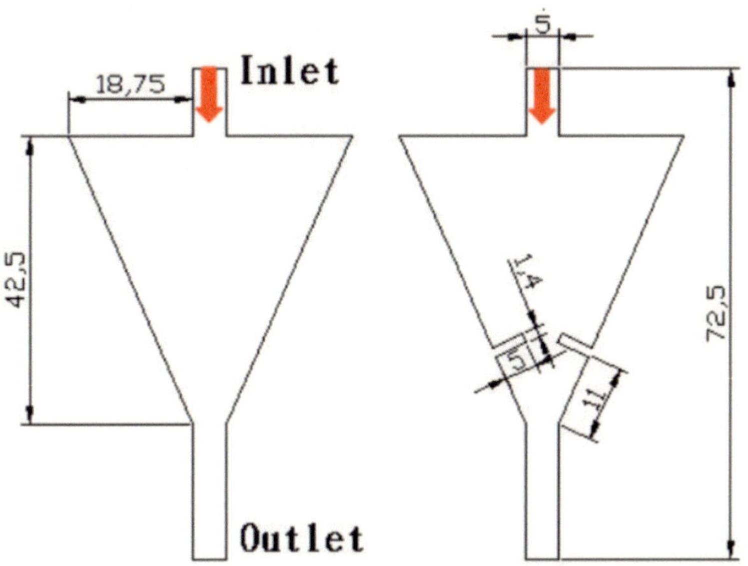

Figure 1. Schematic diagram of convergent flow channel [14].

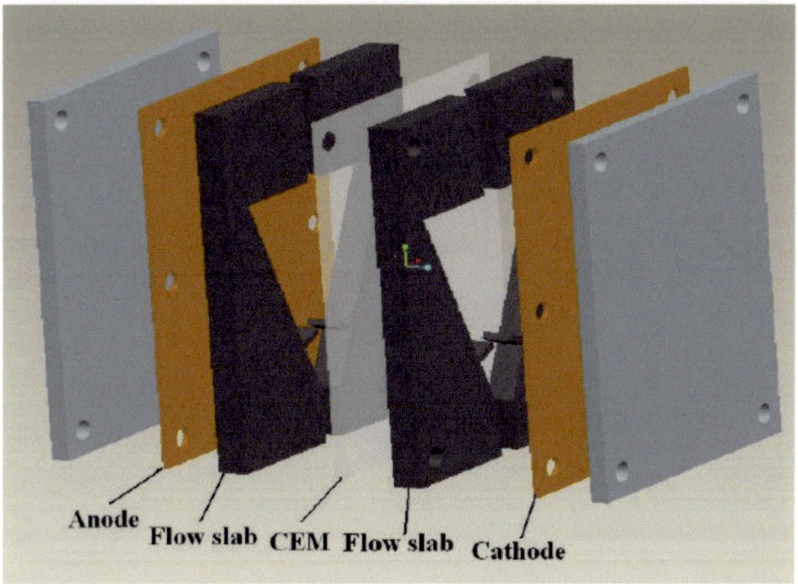

Figure 2. Structure diagram of SMFC [14]

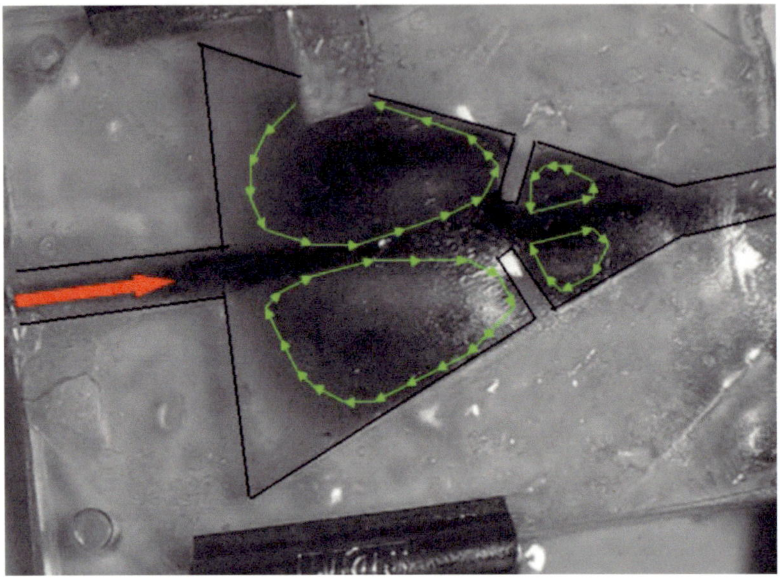

Figure 3. Vortex flow image for convergent flow channel with plates [14]

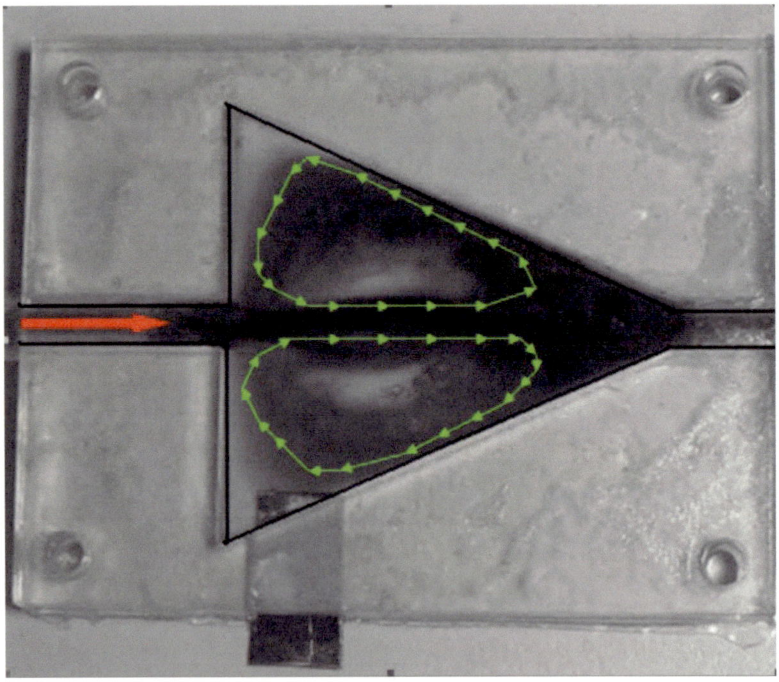

Figure 4. Vortex flow image for convergent flow channel without plates [14]

3.2. Biometric flow channel [15]

Now a biometric concept addressed will be applied in MFCs because the effect of a biometric flow channel (with/without obstacles) on the power generation at different Reynolds numbers (Re=19.85, 496.18), in the single chamber of rumen microbial fuel cells (RMFCs), has been studied by Wang et al.,[15] and the geometrical dimensions and structure of a biometric flow channel can be seen in Figs. 5 and 6.

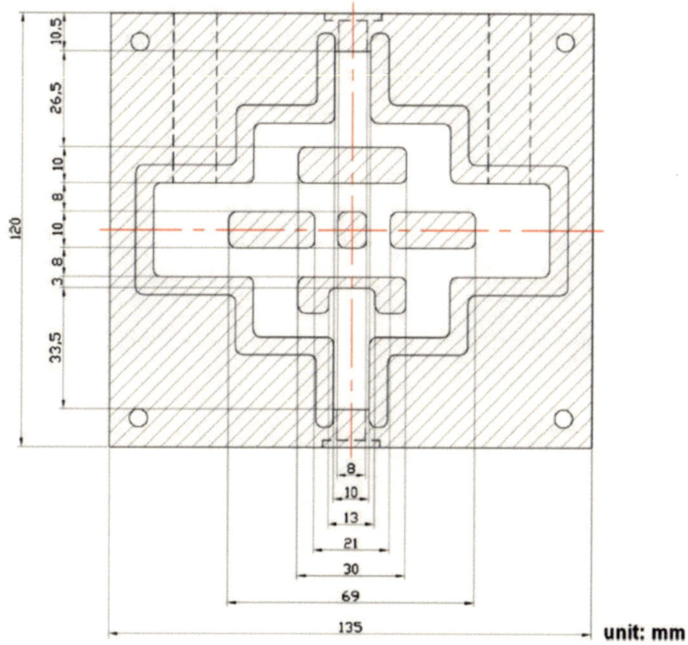

Figure 5. Schematic diagram of biometric flow channel [15].

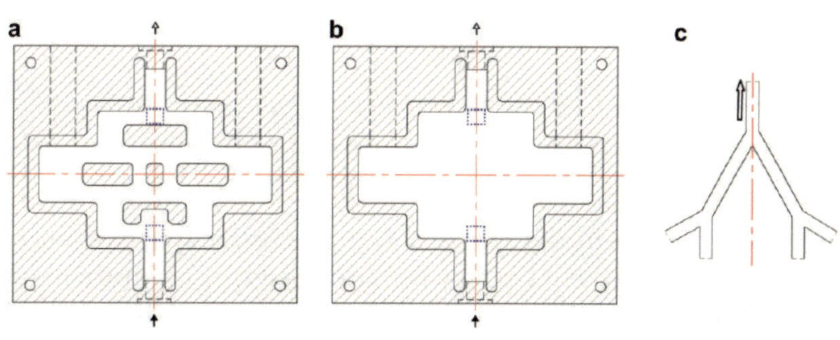

Figure 6. A single chamber (a) with (b) without obstacles marked in the white region for the anode electrode at the bottom of RMFCs [15].

In this study, the simulation result showed that the high flow mixing efficiency obtained by a biometric flow channel with a double two-inlet Y-type inlet channel was above 90%, regardless of being with or without obstacles and different Reynolds numbers. This indicated the mixing mechanism of double two-inlet Y-type inlet channels could greatly enhance the mixing of the reactants before they enter the single chamber of RMFCs.

However, comparing the case of a biometric flow channel without obstacles (No. B/Re=19.85, No. D/Re=496.18), the high Reynolds number [12, 16] and Coanda effect [17] will influence the case of No.D on creating a vortex (as shown in Fig. 7) because it almost did not interact with the main flow and disrupted the reactant atom exchange. Hence, the ohmic and the charge transfer resistance would be increased so that finally some regions of the flow channel based on the need of continuous charging reaction in the fuel cell system would generate a concentration loss. Conversely, the case of No. B would create a smoother flow and a more even reaction at a condition of a lower Reynolds number, thus avoiding concentration loss.

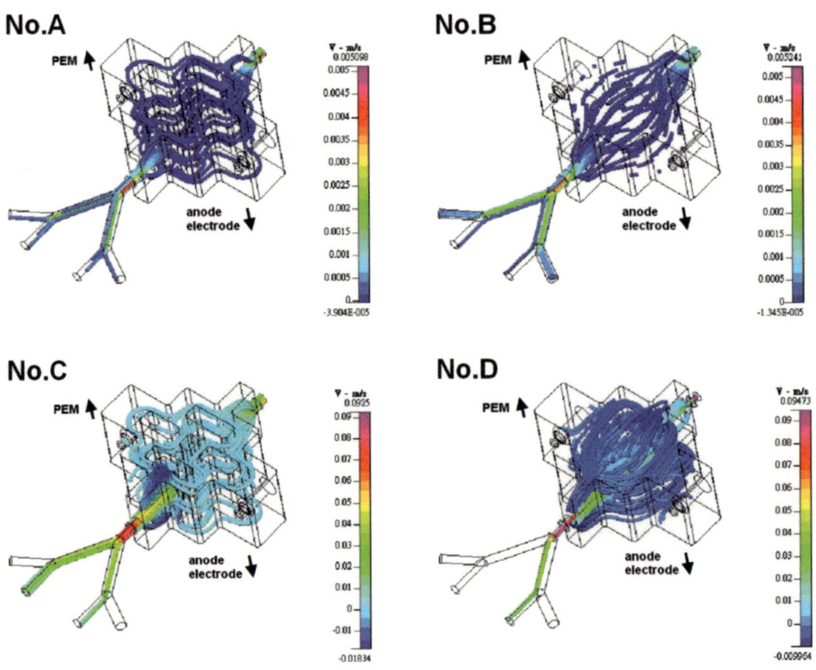

Figure 7. flow velocity images versus different flow conditions and cross-sections analyzed. The location of the anode electrode and proton exchange membranes (PEM) are indicated by arrows [15].

Nevertheless, the flow mixing efficiency of the flow channel in the system was not significantly enhanced by the flow obstacles, but the flow obstacles could evenly distribute the flow field of the reactant [13] on the surfaces of the electrodes and proton exchange membranes. This could avoid concentration loss and also improve power performance [18, 19] along with uneven flow fields. Here, the maximum voltage output and power density in the case of No. C (Re=496.18) would reach 0.716 V and 0.021102 mW/m^2, which is 1.46 times higher than in the case of No. D 1.08. This result indicated that the flow obstacles of No. C would create a more evenly distributed flow field.

On the other hand, the evidence on decreasing the charge transfer resistance for the RMFCs demonstrated that the biometric flow channels with obstacles indeed improved power performance significantly with a higher Reynolds number, despite an increase in the ohmic resistance. The maximum voltage output and power density showed a positive effect at a higher Reynolds number of Re=496.18 by 2.02 times and 4.09 times more than at the lower Reynolds number of Re=19.85, respectively. This is because the obstacles forced the reactant fluid to have an even flow field [20], thus influencing the microbe to be closer to the anode and proton exchange membrane and so made it easier to create a biofilm.

3.3. Biometric flow channel and biometric mixer[21]

The effect of the biometric flow channel (as shown in Fig. 8) and biometric mixer (as shown in Fig. 9) applied to the MFC has been studied by Wang et al. [21]. Wang indicated that adding the biometric flow channel in the anode chamber could enhance the flow mixing efficiency and the homogeneity of the fluid. Therefore, the maximum power density of a MFC reached 91.81 mW/m^3, which is by far greater than that of a MFC without the biometric channel (1.79×10^{-5} mW/m^3).

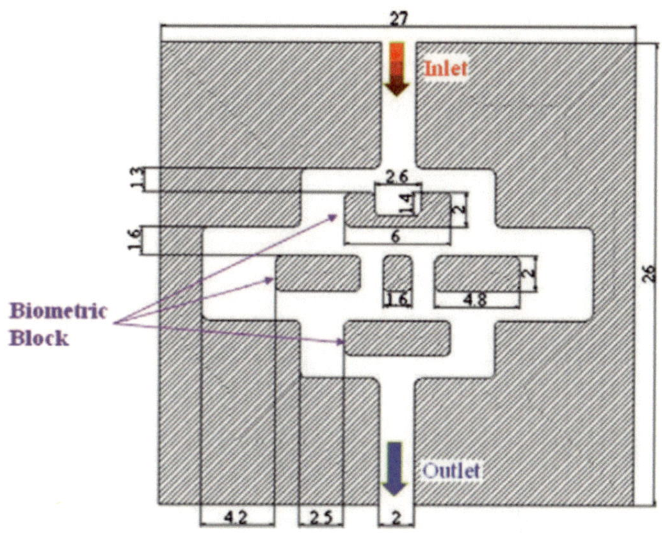

Figure 8. Schematic diagram of the biometric flow channel [21].

The schematic diagram of the biometric flow channel MFC is shown in Fig. 10. However, according to the resulting simulations (as shown in Fig. 11), the biometric mixer generated a high flow mixing efficiency (ε_{mixing}=0.95) of the bacterial liquid and nutrient source before entering the MFC. Then the biometric flow channel enhanced the flow mixing of the reactant

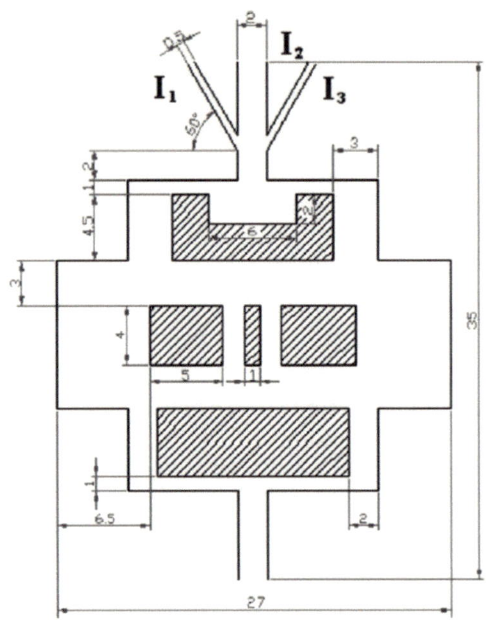

Figure 9. Schematic diagram of the biometric mixer [21].

and a high mixing efficiency of ε_{mixing}=0.98 would be reached at the outlet of the MFC, therefore the maximum power density of the MFC reached 118.34 mW/m^3, which is an improvement of 28.9% compared to the MFC without an added external mixer.

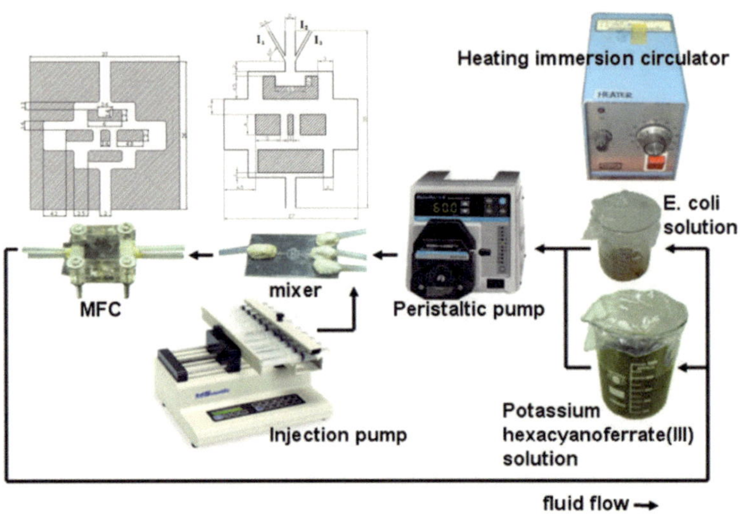

Figure 10. Experimental framework of mixer connected to MFC [21]..

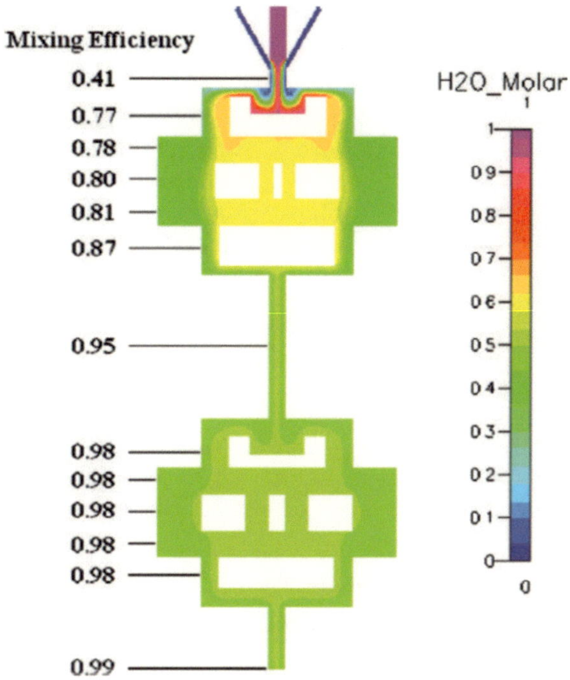

Figure 11. Simulation diagram of mixer connected to a MFC [21]..

3.4. Membraneless microfluidic microbial fuel cell [22]

A membraneless microfluidic microbial fuel cell was used in the Y-shaped microchannel (as shown in Fig. 12), to generate a spontaneous laminar flow on a small scale to separate the cathode and anode to replace the proton exchange membrane [23]. Furthermore, the membraneless microfluidic microbial fuel cell can operate on a short start-up time, with the maximum voltage reached in only 15 min and stable in 45 min at 30 ml/h of flow rate for each stream (as shown in Fig. 12).

As electrical generation of the membraneless μMFC requires a laminar flow with only a little crossover of solute [24]; the Reynolds number (Re) must be less than 10 and the Peclet number (Pe) must be close to 10,000 to maintain the laminar flow and reduce the solute crossover [23]. In this study, the flow rate was increased to 30 ml/h, which reached the maximum open-circuit voltage (OCV) of the membraneless microfluidic microbial fuel cell because the increase in flow rate reduced the diffusion crossover of the solute. Furthermore, the maximum OCV is obtained at Re=8.39 and Pe=7163, which is similar to Choban et al., [23].

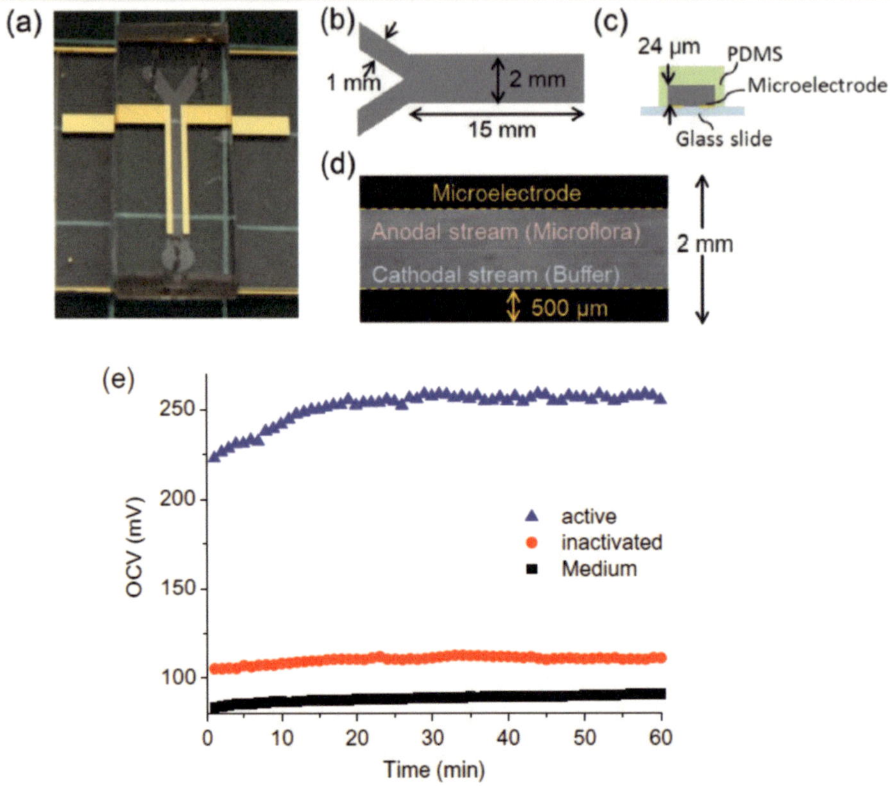

Figure 12. The scheme of the membraneless µMFC and the OCV obtained from the membraneless µMFC [22].

4. Conclusion

Several methods of flow control addressed have been shown to improve the power performance of the microbial fuel cell and can be applied to different applications. A lot of related research has focused on the large scale application, such as wastewater treatment, but a more potential challenge to be considered is a small scale application. The microfluidic microbial fuel cells have the opportunity to be used on biomedical recipients, but the study of the flow control for microbial fuel cells is still limited and there is a need for improvement. Therefore a more sustained research on the method of the flow control, which could be able to improve the practicability of the microbial fuel cell, will be necessary in the future.

Acknowledgements

The author would like to thank the Ph.D candidate of Yan-Ming Chen for his efforts on this chapter.

http://dx.doi.org/10.5772/58346

Author details

Chin-Tsan Wang

National I Lan University, Taiwan

References

[1] Logan BE. Microbial fuel cells. 1st ed., Wiley-Interscience, New Jersey, USA (2008).

[2] He, Z., Minteer, S. D., & Angenent, L. T., Electricity generation from artificial wastewater using an upflow microbial fuel cell, *Environmental science & technology, 39*(14), 5262 (2005).

[3] Kim, H. J., Park, H. S., Hyun, M. S., Chang, I. S., Kim, M., & Kim, B. H., A mediatorless microbial fuel cell using a metal reducing bacterium, *Shewanella putrefaciens, Enzyme and Microbial Technology, 30*(2), 145 (2002).

[4] Lefebvre, O., Tang, Z., Fung, M. P., Chua, D. H., Chang, I. S., & Ng, H. Y., Electrical performance of low cost cathodes prepared by plasma sputtering deposition in microbial fuel cells, *Biosensors and Bioelectronics,31*(1), 164 (2012).

[5] Osman, M. H., Shah, A. A., Wills, R. G. A., & Walsh, F. C., Mathematical modelling of an enzymatic fuel cell with an air-breathing cathode, *Electrochimica Acta, 112*, 386 (2013).

[6] Kim, J. R., Cheng, S., Oh, S. E., & Logan, B. E., Power generation using different cation, anion, and ultrafiltration membranes in microbial fuel cells, *Environmental science & technology, 41*(3), 1004 (2007).

[7] Kjeang, E., Djilali, N., and D. Sinton, Microfluidic Fuel Cells: A Review, *J. Power Sources*, 186, 353 (2009).

[8] Liang, P., Huang, X., Fan, M. Z. Cao, X. X., and C. Wang, Composition and Distribution of Internal Resistance in Three Types of Microbial Fuel Cells, *Appl. Microbiol Biotechnol*,77,551(2007).

[9] Raman, K., and J. C. W. Lan, Performance and kinetic study of photo microbial fuel cells (PMFCs) with different electrode distances, *Applied Energy*, 100, 100 (2012).

[10] Oh, S., Min, B., and B. E. Logan, Cathode Performance as a Factor in Electricity Generation in Microbial Fuel Cells, *Environ. Sci. Technol*, 38, 4900 (2004).

[11] Wang, C. T., Shaw, C., K., and T. Y. Hu, Optimization of Flow in Microbial Fuel Cells:An Investigation into Promoting Micro-Mixer Efficiency with Obstacle, *TKJSE*, 14, 25 (2011).

[12] Trirron DJ. Physical fluid dynamics. 2nd ed., Oxford University Inc, New York, USA (1988).

[13] Wang CT, Hu TC, Hu TY. Biophysical micromixer, *Sensor*, 9, 5379 (2009).

[14] Wang, C. T., Fang, S. J., and Z. Q. Qi, Design of Convergent Flow Channel Applied to Sediment Microbial Fuel Cells, *APPEEC*, 1 (2011).

[15] Wang, C. T., Yang, C. M., Chen, Z. S., and T. Shuai, Effect of Biometric Flow Channel on the Power Generation at Different Reynolds numbers in the Single Chamber of Rumem Microbial Fuel Cell (RMFC), Int J Hydrog Energy, 9242 (2011).

[16] Simpson, R. L., Turbulent boundary-layer separation, *Annual Review of Fluid Mechanics*, 21(1), 205 (1989).

[17] Lashkov, Y. A., Sokolova, I. N., & Shumilkina, E. A., Jet flow over ribbed curved surfaces, *Fluid dynamics*, 27(1), 135 (1992).

[18] Logan, B. E., Hamelers, B., Rozendal, R., Schröder, U., Keller, J., Freguia, S.,... & Rabaey, K., Microbial fuel cells: methodology and technology. *Environmental science & technology*, 40(17), 5181 (2006).

[19] Du Z, Li H, Gu T., A state of the art review on microbial fuel cells: a promising technology for wastewater treatment and bio-energy, *Biotech Adv*, 25464 (2007).

[20] Huang, L., Cheng, S., Rezaei, F., & Logan, B. E. Reducing organic loads in wastewater effluents from paper recycling plants using microbial fuel cells. *Environmental technology*, 30(5), 499 (2009).

[21] Qi, Z. Q., Fan, S. J., Wang, C. T., and Z. Y. Hu, Mixing effect of biometric flow channel in microbial fuel cells, *Appl Energ*, 100, 106 (2012).

[22] Wang, H. Y., & Su, J. Y., Membraneless microfluidic microbial fuel cell for rapid detection of electrochemical activity of microorganism, *Bioresource technology.*,145, 217 (2013).

[23] Choban, E.R., Markoski, L.J., Wieckowski, A., Kenis, P.J.A., Microfluidic fuel cell based on laminar flow, J. Power Sources, 128 (1), 54 (2004).

[24] Sprague, I.B., Byun, D., Dutta, P., Effects of reactant crossover and electrode dimensions on the performance of a microfluidic based laminar flow fuel cell.,Electrochimica Acta, 55 (28), 8579 (2010).

Chapter 2

Biochemical and Electrochemical Perspectives of the Anode of a Microbial Fuel Cell

Aishwarya Mahadevan,
Duminda A. Gunawardena and Sandun Fernando

Additional information is available at the end of the chapter

http://dx.doi.org/10.5772/58755

1. Introduction

1.1. History of biofuel cell

The statement "Perhaps the most refined fuel cell system today is the human body, a mechanism that catalytically burns food (fuel) in an electrolyte to produce energy, some of which is electrical" highlights the connection between living organisms and electricity [1, 2]. With an experiment conducted using frog leg, Biologist Lugi Galvani in 1780's proved that electrical energy and biology have a close connection to each other [3]. Michael Cresse Potter, a Botany professor also demonstrated that living organisms can generate voltage and deliver current [4].

The term "fuel cell" has been in use for over a century. Despite some uncertainty about who first fabricated one, credits of designing and experimenting with first fuel cells go to both Sir William Grove (1839) and the Swiss scientist Christian F. Shoenbein (1868). In early 19th century, different organisms like bacteria, algae, and yeast were considered for this research. With the advent of space race, considerable attention was given to energy generation from recycled waste which in turn ignited interest in microbial fuel cell research. Later, during the sixties and in the early seventies, fuel cell related research accelerated as a consequence of increase in oil prices and has sustained momentum to date. [5, 6]. The time lines of fuel cell development are shown in the Fig.1.

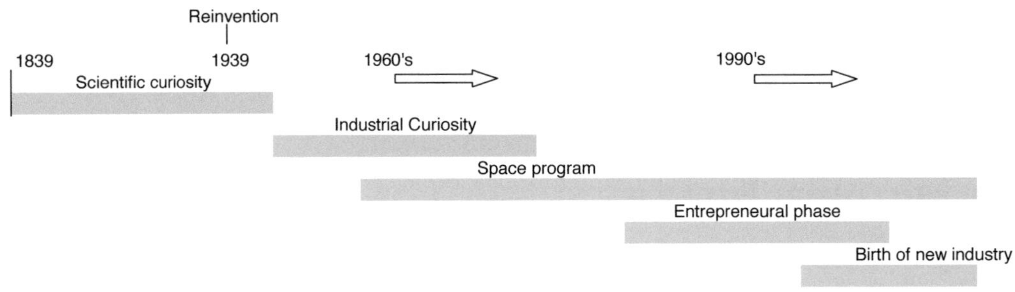

Figure 1. The time line of fuel cell development [7]

1.2. Types of fuel cells

Fuel cells could be broadly categorized into abiotic fuel cells of which the fuel cell components do not comprise any biological material and biotic or biological fuel cells which comprises living organisms or biological material (such as enzymes or derivatives). The primary types of abiotic fuel cells grouped according to the electrolyte used are shown in Table 1.

Type	Features
Alkaline fuel cells (AFC)	Uses KOH as the electrolyte and electro-catalysts such as Ni, Ag and metal oxides
Polymer electrolyte membrane fuel cells (PEMFC)	Uses a proton conductive polymer membrane as the electrolyte and Pt as the catalyst
Phosphoric acid fuel cells (PAFC)	Uses concentrated phosphoric acid as the electrolyte and Pt as the catalyst
Molten carbonate fuel cells (MCFC)	Has a combination of alkali metals (Li, K, or Na)
Solid oxide fuel cells (SOFC)	Uses non-porous metal oxide(s) as the electrolyte

Table 1. Different types of commonly known inorganic fuel cells [7].

The biological fuel cell (BFC) can be categorized into two main areas:

1. Microbial fuel cells (MFC)

2. Enzymatic fuel cells (EFC)

The biological fuel cells (BFC) use enzymes or microorganisms as catalysts. In a microbial fuel cell, the oxidation reactions that are catalyzed by microbes; alternatively, when the catalyst is an enzyme, the cell is called as an enzymatic fuel cell. While both microorganisms and enzymes catalyze oxidative reactions that takes place at the anode, only enzymes (sometime coupled with inorganic catalysts) are used in the cathode. Biological fuel cells utilize organic substrates (such as sugars and alcohols) and operate at mild temperature environments where biological activity is optimal. For example, the catalyst used in a microbial fuel cell could simply be an organism like Baker's yeast that feed on simple sugars or an advanced species like *R. ferrireducens* [8-10] that thrive on more complex substrates.

1.3. Different categories of microbial fuel cells

Various types of MFC designs have been developed of which five main categories are common:

1. **Uncoupled bioreactor MFC**: a separate compartment where organisms produce the hydrogen (fuel) and that hydrogen is fed into a hydrogen fuel cell.

2. **Integrated bioreactor MFC**: hydrogen fuel production and the electricity generation both take place in the same chamber.

3. **MFC with mediated electron transfer**: where intermediate molecules shuttle electrons from microbial cells to the electrode.

4. **MFC with direct electron transfer**: where electron transfer to the electrode take place without the presence of any mediator molecules.

5. **Mediator-less and membrane-less microfluidic MFC**: an emerging type of MFC that eliminates use of mediators and cation exchange membrane.

The first two designs use hydrogen that is biologically generated and uses this hydrogen in a PEM-like fuel cell system. Thus, these fuel cells possess similarities with their inorganic counterparts. There are different types of bacteria and algae that generate hydrogen under anaerobic conditions, e.g. *Escherichia coli, Enterobactor and C. butyricum*that can be used in such reactors. Theoretical output in these fuel cells is in the neighborhood of 10% but in actual environment, the values have been slightly less. Enzyme interactions (such as hydrogenases) with H_2 is one reason for performance reduction. The inhibition of anaerobic hydrogenases by evolving oxygen (during photosynthesis) is another. Contamination of H_2 with other gaseous species such as CO and H_2S is also considered to contribute toward inefficiencies. The second design mentioned above has similar disadvantages to the first with the only difference being the position of the bioreactor [11, 12]. Such MFCs commonly use Pt as electrode catalysts.

In the third type, an intermediate molecule known as mediator is used for electron transport [13]. The mediator will shuttle electrons between the electrode and redox enzyme in the microorganism [14] following redox cycles as given in Equations1&2. Different mediators that have been utilized in microbial fuel cells are listed in Fig.2.

$$C_6H_{12}O_6 + \text{Mediator}_{(o)} \longrightarrow \text{Product} + \text{Mediator}_{(r)} \tag{1}$$

$$\text{Mediator}_{(r)} \longrightarrow \text{Mediator}_{(o)} + \text{Electrons} \tag{2}$$

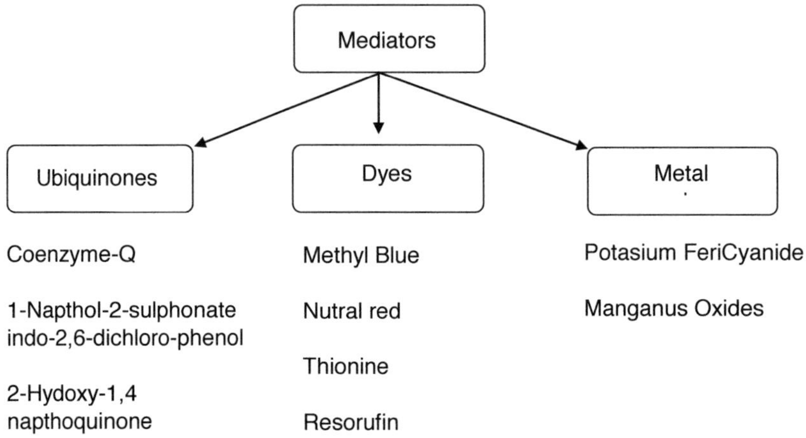

Figure 2. Different mediators used in biological fuel cells.

The characteristics that the mediators must possess are well established [15, 16]: 1) The molecules should be able to form redox couples; 2) They should be stable in both, reduced and oxidized form; 3) They should not be biologically degradable; and 4) They should not be toxic to the biological species. However, in practice, mediators have commonly contributed to fuel cell performance issues due to degradation and toxicity to the biological medium [17, 18].

The fourth type of MFCs do not contain any extraneous mediators and in this type, bacteria are believed to communicate directly with the electrode using self-made mediators. Studies supporting this hypothesis have been conducted with iron-reducing bacteria such as *Shewanella putrefaciens*, *Geobactor sulferreducens*, and *Rhodeferax ferrireducens* [19]. The fifth type involves miniaturization of MFC using microfluidic technology which is capable of achieving high energy efficiency and durability. This type enables an advantage over conventional MFCs by eliminating the need for membranes (PEM) as a result of co-laminar flow of fuel and oxidant streams that extemporaneously separate anode and cathode in the cell. Electron transport to the anode can occur using electron mediators, or by direct membrane associated electron transfer, or by purported nanowires produced by certain microbes [20, 21].

2. Biochemical and electrochemical phenomena that occur at the electrodes of MFCs

Just like in any other fuel cell, the oxidation reaction in a microbial fuel cell occurs at the anode releasing electrons to and protons at the anode. The concomitant reduction half reaction takes place at the cathode collecting the electrons that travelled through an external circuit and combining them along with the protons with the terminal electron acceptor-which in most cases is oxygen. In order to keep the fuel from being crossed over to the cathode compartment or oxygen from entering the anode compartment (commonly referred to as cross over reactions), a membrane (commonly referred to as proton exchange membrane or PEM) that is selective only for protons (transport) is generally used. A schematic of the reactions that take place in a microbial fuel cell is depicted in Fig. 3.

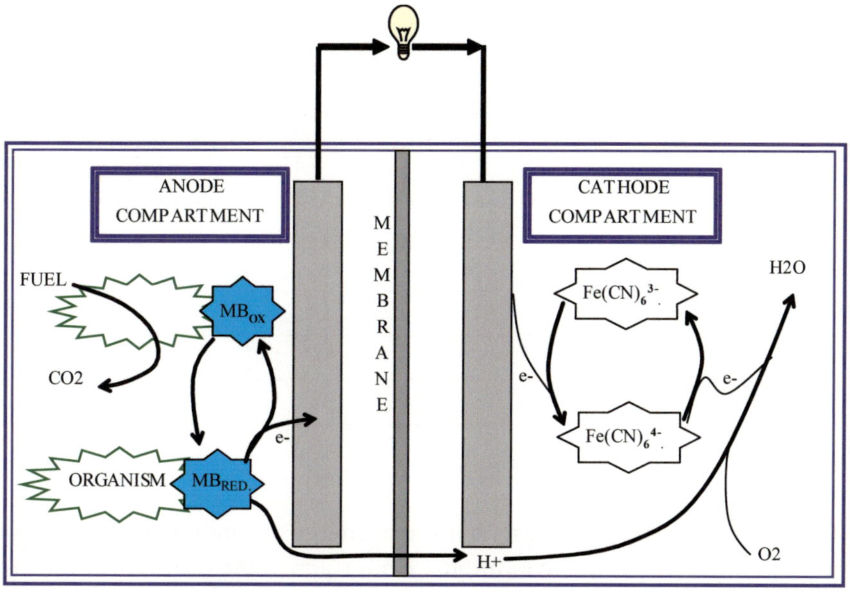

Figure 3. Schematic diagram of a microbial fuel cell and its operation [22]

2.1. Catabolic pathways involved in energy production from microbes

Microorganisms derive energy for living from the free energy produced by fuel oxidation. A part of the free energy produced is retained by the microbes for their catabolic activities and the rest may be utilized to generate electricity. Hence, for sustainable operation of the MFC, it is necessary for the microorganism to balance out the amount of energy consumed for its vitality and that being converted into electrical energy.

There are two metabolic pathways used for energy conversion in microbes namely, respiration and fermentation. In the respiratory or oxidative pathway, the Gibbs free energy is utilized by the microbes for their respiration where the electrons circulate around a respiratory chain and finally exits the microbial cell via membrane-bound electron acceptors.

Under aerobic conditions, oxidation of the fuel glucose follows four distinctive steps [23]: 1) Glycolysis, 2) Krebs Cycle, 3) Electron transport chain, and 4) Oxidative phosphorylation. In glycolysis which is also known as Embden-Mayarhoff-Parnas pathway, the six carbon glucose atom is broken down into two molecules of pyruvate as shown in Equation 3. This takes place in ten successive steps, each of which is catalyzed by a specific enzyme. Pyruvate, the product of glycolysis goes further through a three stage process where it is finally converted to CO_2 and more energy in the form of ATP. The total ATP production from glucose in a respiring system is given in the Table 2.

$$C_6H_{12}O_6 + 2NAD^+ + 2ADP + 2Pi \longrightarrow 2NADH + 2ATP + 2Pyruvate + 2H^+ + 2H_2O \tag{3}$$

Reaction Sequence	ATP Yield
Glucose → Fructose 1,6-diphoshate	-2
2 Trios Phosphate → 2,3-phosphoglyceric acid	2
$2NAD^+ \rightarrow 2NADH \rightarrow 2NAD^+$	6
2Phosphoenol pyruvic acid → 2Pyruvic acid	2
2Pyruvic acid → 2Acetyl Co A + $2CO_2$	
$2NAD^+ \rightarrow 2NADH \rightarrow 2NAD^+$	6
2Acetyl CoA → $4CO_2$	24
Net: $C_6H_{12}O_6 + 6O_2 \rightarrow 6CO_2 + 6H_2O$	38

Table 2. ATP yield for corresponding biochemical reaction(s)

Even though aerobic respiration is the main form of energy generation in organisms, some can generate energy under anaerobic conditions. Under anaerobic conditions, glucose will be converted to pyruvate as it was under aerobic conditions, but in the next stage instead of the Krebs cycle a different pathway will follow-specific to the type of organism viz. prokaryote/eukaryote. While yeast for instance undergoes alcohol fermentation, some organisms follow the lactic acid fermentation.

In microbial catabolism under anaerobic conditions, carbohydrates are oxidized without the presence of oxygen. This oxidation reaction gets partially completed inside the microbial cellular structure and with the help of co-enzymes, the electrons and protons are transferred outside the cellular membrane. The overall reaction of a fermentative MFC is given in Equation 4. This reaction can be broken down to half reactions as shown by Equation 5&6. Inside the anaerobic compartment of a MFC, the reaction represented in Equation 4 takes place where the electrons released will be transferred to anode and the protons (H^+) travel to the cathode compartment (sometimes via a proton exchange membrane). The released electrons from the anode will go through an external load as shown in Fig.3. and consequently enter the cathode.

$$C_6H_{12}O_6 + 6O_2 \longrightarrow 6CO_2 + 6H_2O \tag{4}$$

$$C_6H_{12}O_6 + 6H_2O \longrightarrow 6CO_2 + 24H^+ + 24e^- \ \mathbf{E^0 = 0.014V} \tag{5}$$

$$O_2 + 24H^+ + 24e^- \longrightarrow 12H_2O \ \mathbf{E^0 = 1.23V} \quad (6)$$

Since ATP production is quite low under anaerobic conditions i.e. almost two molecules per one pyruvate molecule, the organisms need rapid processing of $NAD^+$and NADH. For the purpose of harvesting H^+ what matters for a MFC is the amount of co-enzyme (NAD^+) getting involved in the reaction. In that perspective, anaerobic conditions are much favorable for the harvesting of protons when compared with aerobic conditions. Under anaerobic conditions, the rate of NAD^+ to NADH conversion takes place at a significant rate, while under aerobic conditions the already present oxygen in the media reacts with H^+ making it difficult to run the half reaction given by Equation 6.

In a MFC system, the main electron acceptor that is engaged in harboring the electrons released from microbes is a solid anode. The formation of a biofilm on the anode is believed to be driven by the consumption of greater amounts of energy (released by microbial catabolism) by the solid anode when compared to the other electron acceptors. The bridging concepts of electron transfer between microbes and the solid anode is still under speculation and is discussed briefly in the next section.

2.2. Electron transfer mechanisms

The electron transfer mechanism involves amalgamation of knowledge from electrochemistry, biochemistry and microbiology. Electrons pass on to the electrode of MFC as a result of several irreversible enzymatic reactions and eventually the reversible electrochemical reactions of the electron transport chain [24]. For the reversible interfacial reaction to take place at the anode surface, electrons rely on shuttles such as cytochromes, proteins (such as PQQ), bound or soluble redox mediators to reach atleast within 10 Å proximity of the anode [25].

Several electron transfer mechanisms and processes have been proposed including direct electron transfer, indirect electron transfer and shuttle mediated transfer of electrons. Of the various mechanisms proposed, direct contact mechanism indicates formation of monolayer of microbes on the anode surface enabling direct transfer of electrons to the anode via cell membrane or a membrane organelle. Kim et al. [26] observed that certain Fe(III)-reducing bacteria (*Shewanella putrefaciens*) can transfer electrons to the electrode without the aid of synthetic mediators as shown in Fig.4a. Cytochrome, a redox protein perhaps present on the outer membrane of the cell supposedly causes direct electron transfer by its electrochemical activity of reducing water soluble Fe(III). This type of transfer is capable of producing lowest extracellular potential losses due to the negligible gap between microbe and the electrode. However, the electron transfer in this mechanism is limited by the total number of bacteria in direct contact with anode [27]. Also, the c-type cytochrome may not be present in all the Fe(III)-reducing bacteria which further limits the electron transfer to a very few species [28].

Apart from ferric ion reducing bacteria, yeast cells (*Hansenula anomala*) have also been successfully found to enable a direct electrical communication between the cells and anode surface. In this case, electron transfer takes place without any external mediator or any intermediate redox reaction such as Fe(III) reduction, but with the help of redox enzymes,

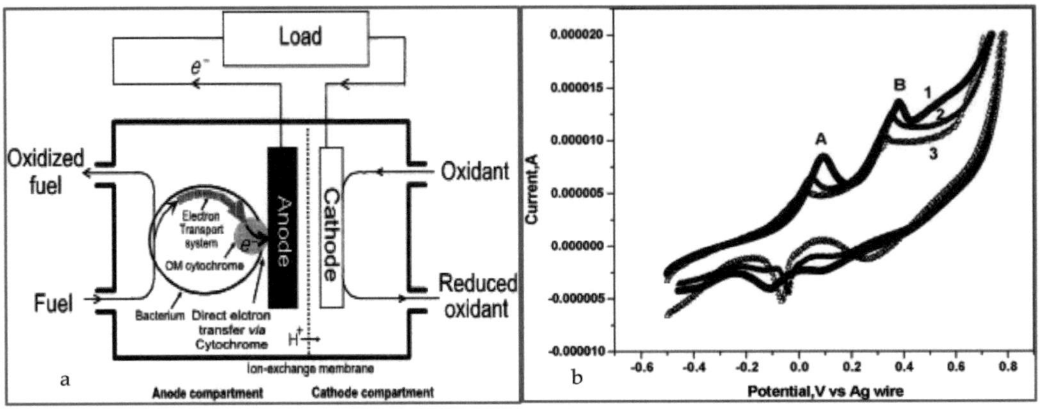

Figure 4. a) Schematic diagram describing direct electron transfer mechanism in MFC [26]; b) CV of yeast biofilm on anode in phosphate buffer at scan rate 50 mV/s. (1) Blank CV; (2 and 3) after successive additions of 0.5 ml each of lactate stock solution [29].

ferricyanide reductase and lactate dehydrogenase that are present on the outer membrane of the yeast cells. As can be seen in Fig.4b., the CV reveals two redox peaks, A and B of the respective redox enzymes that are responsible for the electrochemical activity of the organism. These enzymes get reduced by consuming the electrons that are liberated during the oxidation of the substrate and transfer the electrons to the anode. The CV curves 2 and 3 depict reduction in peak currents upon subsequent addition of substrate indicating direct communication between the enzymes and anode [29].

A recent discovery of interest in the milieu of direct electron transport (DET) mechanism in MFC is the pili growth seen in bacteria (*Geobacter sulfurreducens*). Pili also known as bacterial nanowires, are minute flexible structures made of protein called pillin that help the bacteria to cling to surfaces and distinguish materials in its surroundings. These pili overcome the limitation of the former mechanism by allowing charge transport through multiple layers of biofilms on the anode surface by not only being electrically conductive but also supposedly forming internal networks (see Fig.5.).

The subunit protein, PilA present in pili consists of a cluster of aromatic acids that are believed to play a major role in their conductivity [31, 32]. *Shewanella oneidensis* uses both types of DET mechanisms involving c-type cytochrome and conductive pilus-like nanowires for electron transfer [33, 34]. Conductivity of pili has may be measured using relation (7) given by Malvankar et al. [30]

$$\sigma = G\left(\frac{2a}{gL}\right) \tag{7}$$

where G is the biofilm conductance, L is electrode length, g is thickness of biofilm and a is the non-conducting gap width between two working electrodes (anode).

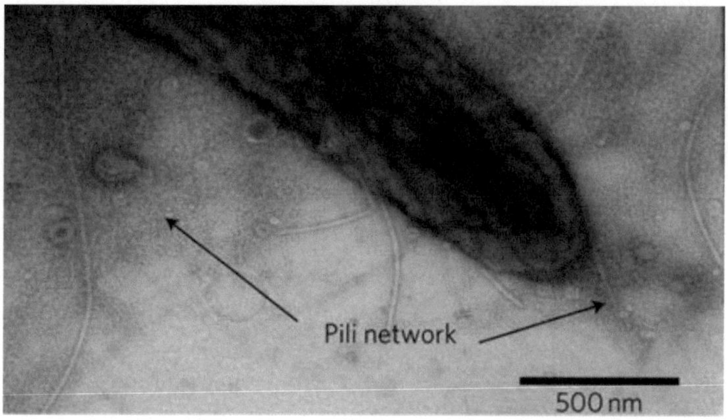

Figure 5. TEM image showing pili formed by the strain KN400 forming an interpenetrating network on the anode biofilm [30]

While there have been positive findings on bacterial nanowires toward long-range electron transfer on anode; further in-depth investigations are required to find precisely what feature of the pili facilitates metal-like conductivity in pili and other microbial species that possess this trait [35].

Although direct contact mechanism is much sought after for electron transfer, extensive research on indirect or mediated electron transfer mechanism indicates that it is a conventional and efficient method for current generation in MFCs [36]. Indirect electron transfer mechanisms can be categorized based on the type of mediator used to wire the microbial catabolism with anode surface. A good mediator displays high membrane permeability, it is adequately soluble, possesses high electron transfer rate, non-toxic to microbial cells and non-biodegradable. In theory, mediators with low redox potentials (such as SO_4^{2-}) are said to be favorable for MFCs. This is because, electrons always flow from the low redox potential (of mediator) to the high redox potential (of electron acceptor or anode). However, mediators with high redox potentials have superior affinity for drawing electrons from the electron carriers in the cell.

Initially artificial redox mediators such as phenazines, phenothiazines, phenoxazines and quinones were used to carry out the electron transfer in the MFC systems. However, there are several disadvantages associated with these mediators that makes their use unsustainable and impractical – low current densities, difficulty in long range electron transfer over thick biofilms, need for regular addition of fresh mediators and toxicity. Hence, it was necessary to employ the intrinsic production of metabolites in microbes for electron transport purposes.

Microorganisms are capable of producing endogenous metabolites using primary and secondary pathways to carryout various biological processes. Primary metabolites such as H_2 [37] and H_2S [38] that are produced by microbial catabolic oxidation of fuel (anaerobic respiration and fermentation) have also been successfully used as redox mediators. Some secondary metabolites that have been considered for use in MFC applications are phenazine-1-carboxamide, pyocyanine (*Pseudomonas aeruginosa*) [39], neutral red, anthraquinone-2,6-disulfonate (AQDS), thionine, methyl viologen, methyl blue, humic acid [40] and 2-amino-3-

carboxy-1,4-naphthoquinone (ACNQ) (*Bifidobacterium longum*) [41]. These mediators display cyclic redox behavior, which means that a single molecule can be used continuously for electron transfer to and fro the anode or biofilm. This is not only sustainable but also convenient for long range electron transfer in the anodic biofilms resulting in constant and enhanced generation of current.

3. Factors that affect performance of microbial fuel cells and potential remedies

3.1. Polarization losses

Theoretically, MFC can attain a maximum cell voltage (emf) of 1.1 V under open circuit conditions. However, in reality it undergoes numerous forms of (polarization) losses and can achieve only 0.8 V [42] in open circuit conditions and around 0.62 V [43] during current generation. Overvoltage which is the difference between the theoretical and measured cell voltage collectively represents the overpotentials of the electrodes as wells as the overall ohmic loss of the system [32]. Prominent sources of these overpotentials are the intracellular and extracellular potential losses that occur in the biofilm formed on anode.When the internal resistance is high, there is a significant loss of charge in the system, thus reducing the effective voltage available at the end terminal [7, 44]. Key types of polarization losses (Fig.6.) viz. activation polarization, concentration polarization, ohmic losses and losses due to microbial metabolic activities are discussed in more detail below.

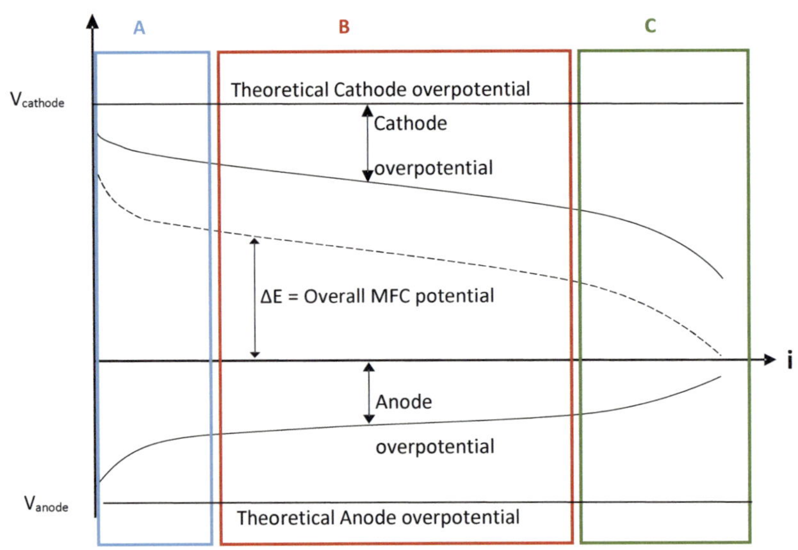

Figure 6. The polarization curve for a typical MFC

Activation losses

Usually, to initiate transport of charges liberated by fuel (electron donor) oxidation to the anode (electron acceptor), an energy barrier must be overcome. The energy barrier comprises of an additional potential known as activation overpotential that is required for transfer of electrons from fuel to microbial shuttles and finally to the anode. Activation losses, represented by region A in Fig.6. is observed as a sharp decrease in MFC voltage (low polarization) at the initial low current densities, but it is steadily overshadowed by ohmic and concentration losses which usually occur at intermediate or high current densities (region B in Fig. 6). Clearly, activation losses occur at both anode and the cathode, and it is important to note that the cathodic overpotentials are much larger than anodic overpotentials. With the increased exchange current density this overpotential can be reduced [7, 44]. The activation losses can be explained from the Tafel equation shown by Equation 8.

$$\Delta \mathrm{Vact} = \mathrm{A} \times Log(\frac{i}{i_0}) \tag{8}$$

where ΔVact: activation overpotential, A: Tafel Slope, i: current density, i_o: exchange current density.

Some steps that can be taken to minimize the activation losses are:

- **Increasing anode surface area**

Increasing the surface area is a reliable approach to decrease the activation potential as when the surface area is increased the current density gets reduced. This can be done by increasing the electrode surface porosity and roughness.

- **Improving anode-microbe interactions**

In order to decrease the activation losses at the bacteria, it is necessary to improve anode-microbe interactions. Using the correct mediator would eliminate this problem by enhancing electron transfer. As mediators would go inside the cell membrane it can reduce the intracellular activation losses as well. MFC systems employing microorganisms that produce conducting pili have relatively low activation polarization.

- **Increasing the operating temperature**

In an inorganic fuel cell raising the temperature would reduce the activation overpotential but in the microbial fuel cell or in the enzymatic fuel cell it is not possible to increase the temperature unless the bio-reaction section is separate from the anode chamber.

- **Decreasing the activation loss at the electrode surface**

The activation energy at the electrode surface can be decreased by adding catalyst to the electrode. The catalysts that has been widely tested is Pt, which is reported to get polluted by bacterial suspensions. It has been reported that some success has been achieved by coating the electrode with a conducting layer that shield microbes from direct contact with the catalyst material [45]. Immobilization of catalysts such as neutral red (that also acts as a mediator) and manganese oxide on the electrode surface have shown to increase MFC power output. [15, 46]

The loss of potential due to internal current and crossover of reactants would also be significant if the fuel cell is operating at low current densities. Some electrons will pass through membrane rather than through the external circuit. The membrane is impermeable to the oxygen molecule but certain percentage can be diffused into the anode where it reduces the current that can pass through the external circuit.

Ohmic losses

Ohmic losses depicted as the medium polarization region B in Fig.6. is caused when the flow of charge is hindered as a result of the anodic resistance. The potential drop can be easily represented by ohms law as given in Equation 9 where Ri represent the total internal resistance and I is the circuit current.

$$\Delta V = I \times Ri \tag{9}$$

Increasing the conductivity of anode material, minimizing contact resistance and the total travel distance of electrons within the anode helps in limiting the ohmic losses. Use of highly conductive anode materials with 3D architecture (eg. 3D graphite felt electrode) has shown to produce higher current generation by overcoming the ohmic losses [47]. The three dimensional structure not only offers a high surface to volume ratio but also an evident increase in the anode-microbe interaction, thus facilitating higher electron transport [48]. Apart from this, resistance caused by the internal connections in the MFC system and the cation exchange membrane against the ionic flux also contributes to ohmic losses. The anodic (electrical) resistance were reported to be negligible when graphite electrodes were used [7] and the contact resistance can also be significantly low as compared to the ionic resistance. Optimizing the electrode spacing, using a low resistance membrane while improving the conductivity and buffer capacity of the electrolyte (tolerable by the microorganism) are concomitant strategies to improve ion transfer through the membrane [49, 50].

Concentration losses

Imbalance in rate of mass transfer of substrate and products to and from the anode respectively and the total current generated in the system may result in increase in anode potential and decrease in cathode potential or vice versa causing concentration (or mass transport) losses. These losses are most prominent at high current densities due to diffusion-limited mass transfer of fuel to the anode surface. Also, the accumulation of oxidized products and cations in the biofilm may change the redox conditions and alter the metabolic activities of the microbes. Hindrance in cation transport may further cause a pH gradient between the electrodes leading to a significant reduction in the power output. Anode design and operational parameters are contributing factors toward concentration losses represented by the maximum polarization region C in Fig.6.

Losses due to fuel scavenging metabolic processes of microrganisms

Loss of voltage can also occur due to catabolic activities of the microbe while deriving energy from fuel oxidation. As discussed earlier in the chapter, in an effective MFC, the anode potential should be as low as possible to allow attainment of high MFC voltage and adequate catabolic energy gain for the survival of microorganisms. Nevertheless, extremely low anode potential can hinder electron transfer causing fermentation of fuel while producing high energy products, resulting in loss of electrons. Furthermore, this also leads to added electron losses by excessive buildup of anodophilic biomass. A number of factors such as type of microbes, community composition, anode-microbe interaction, rate of fuel degradation by the microbes, number of microbes actively degrading the fuel and mix up of fuel through the electrolyte between the electrodes can affect the microbial metabolic losses.

3.2. Microbial interaction with the anode surface

The electrical performance of a MFC is largely dependent on how well the microorganisms interact with the anode. A prime requirement here is that the biofilm that comprise of the microbes is adhered properly onto the anode.

Microbial adhesion on anode surface can be understood by the notion of surface charges. Most of the micro-organisms are negatively charged by nature and hence, attract positively charged surfaces. So, several (surface) modification techniques have been employed to facilitate this charge attraction process. For example, treating the anode surface with ammonia has been successfully attempted. Ammonia treatment facilitates the negatively charged bacteria to readily attach to the now positively charged anodes. The power output of the treated anode was expectedly much higher than its non-treated counterpart. However, high temperature requirements and, complex conditions and instrumentation have made this process commercially less feasible. In a different instance, treatment of activated carbon felt anodes with nitric acid (acid treatment) for rendering positively charged surfaces has resulted in a 58% increase in power density [51]. Zhou et al. reported that electrochemical oxidation of anode led to the change of the anode properties, such as augmented surface area, reduced internal resistance and anode potential, and therefore aided to the microbial adhesion and electron transfer on the anode surface [52].

Role of anode material and architecture

Anode material and its architecture directly affects microbial adhesion, electron transfer and fuel oxidation. The noble metal electrodes (Pt, Au, Ag, Pd) have been reported to be less attractive as MFC anodes due to their high cost and weak adhesion of microbes; and other high-performing electrodes (Ni, Cu, Rh, Ir) are being sought as alternatives [53].

Carbon-based anodes have been extensively tested for MFC applications as they exhibit superlative properties such as high conductivity, durability, eco-friendliness and their flexibility to be shaped into various architectures. They can have planar, packed or brush like configuration(s). The conventional electrodes include graphite - rod, felt, plate and fiber brush, carbon - felt, cloth and paper, and reticulated vitreous carbon (RVC) [54, 55]. Logan et al. made an observation that packed and brush design of anodes gave higher power output than the planar type anodes by increasing the anode specific surface area and consequently the volumetric density of exoelectrogenic bacteria [55]. Additionally, the brush configuration also has high porosity. The fibrils of the brush anode helps the micro-organisms to hold onto the anode structure. Due to its enhanced conductivity and non-corrosive nature, titanium was found to be a suitable core wire on which the carbon or graphite fibers are wound.

Lately, nanomaterials have attracted much attention in various fields due to their unusual yet beneficial structural, chemical and electrical properties. Use of nanomaterial for anodes and electron transport has been reported to augment the performance of MFCs. Fan et al. showed that Au and Pd nanoparticle decorated anodes produced enhanced current densities than that of the control electrodes. They also observed that the anodic performance was significantly affected by not only the chemical composition of nanoparticles but also their size and shape [56]. In 2012, Xu et al. tested Fe nanoparticle-decorated graphite disks which resulted in approximately six-fold higher average current densities than the plain graphite anode. Upon running a whole genome microarray analysis of the gene expression of *Shewanella oneidensis* used in this study, they found out that genes encoding biofilm formation were significantly up-regulated as a response to nanoparticle-decoration [57]. However, it should also be noted that the probable cytotoxicity of certain nanoparticles often limit the possibilities that can be reached using nano-materials for MFC applications.

Coating the anode surface with materials such as carbon nanotubes (CNTs), conductive polymers and nanopolymer composites are other methods used for anode surface modification. Conductive polymer based anodes utilizing polypyrrole [58], poly(3-hydroxy butyrate-co-3-hydroxyvalerate) [59], nanowire networks [60] and nano/-composites of the polymers [61, 62, 63] have been studied and found to significantly increase MFC performance.

Recent studies show that CNTs are promising electrode materials due their high surface area, superior electrical conductivity, chemical inertness, decreased startup time and low internal resistance [64]. Functionalizing the CNTs appropriately, may further enhance electron transport. Mink et al. [65] made observations on a forest type multi-walled carbon nanotubes (MWCNTs) with a nickel silicide contact area that produced current density of 197 mA/m^2 and power density of 392 mW/m^3. MWCNTs were said to have increased the anode surface-to-volume ratio, which improved the ability of the microbes to couple and transfer electrons to the anode. Nickel silicide were reported to boost the output current by providing a low resistance contact area that allowed efficient shuttling of electrons.

A three dimensional architecture that facilitates augmented growth of the microbes can be achieved by embedding CNTs on carbon cloth and polyester fabric using doping techniques. Also, the CNT coating stimulates active surface interactions with the microbes enabling direct electron transfer thereby giving 68% high power density and 10-fold-lower charge-transfer resistance than the traditional carbon cloth based anodes [48]. Yet another advancement in the three dimensional anodes was the 3D conducting graphene–polyaniline framework that was reported to outperform the planar carbon electrode by additionally providing multiplexed and highly conductive pathways [66]. Wang et al. [67] developed a 3D reduced graphene oxide–nickel foam as an anode for MFC which achieved a remarkable volumetric power density of

http://dx.doi.org/10.5772/58755

661 W/m^3 based on volume of the anode, which is the highest value yet obtained for a MFCs with a pure strain of *S. oneidensis* MR-1.

4. Conclusion

In summary, MFCs have tremendous potential to generate electrical energy from chemical energy present in organic fuels primarily for decentralized stationary power generation applications. However, challenges associated with effective mass and charge transport along with the intricacies associated with making the living organisms interact with the inorganic (electrode) world have hampered this technology coming to technological fruition as yet. Advances in material science coupled with nanotechnology may provide novel tools to effectively harvest, transport and utilize electrical charges generated by MFCs for useful applications in the forthcoming future.

Author details

Aishwarya Mahadevan, Duminda A. Gunawardena and Sandun Fernando*

Biological and Agricultural Engineering Department, C Scoates Hall, Texas A&M University, College Station, USA

References

[1] Keneth, L. Biochemical Fuel Cells. in Symposium on Bioelectrochemistry of Microorganisms. 1966.

[2] Mitchell, W., Fuel cells. 1963: Academic press Inc. NewYork.

[3] Cole, K.S., Membranes, Ions, and Impulses. 1972.

[4] Potter, M.C., Bacteria as Agents in the Oxidation of Amorphous Carbon. 1910.

[5] Cohen, B., The bacterial culture as an electrical half-cellJournal of Bacteriology, 1931.

[6] Wingard, L., et al., Bioelectrochemical fuel cells Enzyme Microb. Technol,, 1982.

[7] Barbir, F., Pem Fuel Cells theory and practice. 2005: Elsevier.

[8] Bennetto, H.P., Electricyt generation by Microorganism. Biotechnology Education, 1990. 1: p. 163-168.

[9] Bennetto, H.P., Microbial Fuel Cells Life Chemistry Report. 1984. 2: p. 363-453.

[10] Aston, W.J. and T.A.P. F, Biosensors and Biofuel cells. Biotechnology General Engineering Review, 1984: p. 89-120.

[11] Cooney, M.J., R. E, and M.I. W, Physiologic Studies with the sulfate-reducing bacterium Desulfovibrio desulfuricans. Enzyme Microb. Technol,, 1996. 18(5): p. 358-365.

[12] Scholz, F. and U. Schroder, Bacterial Batteries. Nature Biotechnology, 2003. 21(10): p. 1151-1152.

[13] Tender, L.M., et al., Harnesssing microbially generated power on the seafloor. Natural Biotechnology, 2002. 20(8): p. 821-825.

[14] Roller, S.D., et al., Electron transfer coupling in microbial fuel-cells 1. Comparison of redox-mediator reduction rates and respiratory rates of bacteria. J. Chemical Technology Biotechnology, 1984. 34B: p. 3-12.

[15] Park, D.H. and J.G. Zeikus, Improved fuel cell and electrode designs for producing electricity from microbial degradation. Bioenergy and Biotechnology, 2003. 81(3): p. 348-355.

[16] Willner, I., et al., Biofuel cell based on glucose oxidase and microperoxidase-11 monolayer functionalized electrode. J. Chemical Soc.-Perkin Trans., 1998. 2(8): p. 1817-1822.

[17] Delaney, G.M., Bennetto, H. P., Mason, J. R., Roller, S. D., Stirling, J. L. and Thurston, C. F., Electron-transfer coupling in microbial fuel-cells 2. Performance of fuel-cells containing selected microorganism mediator substrate combinations. J. Chem. Technol. Biotechnol., 198434(1): p. 13-27.

[18] Gil, G.C., et al., Operational parameters affecting the performance of a mediator-less microbial fuel cell.Biosens. Bioelectron., 2003. 18(4): p. 327-334.

[19] Chaudhuri, S.K.a.L., D. R., Electricity generation by direct oxidation of glucose in mediatorless microbial fuel cells. Nat. Biotechnol. vol. 21: pp.1229-1232.. Natural Biotechnology, 2003. 21: p. 1229-1232.

[20] Wang, H.-Y. and J.-Y. Su, Membraneless microfluidic microbial fuel cell for rapid detection of electrochemical activity of microorganism. Bioresource Technology, 2013. 145(0): p. 271-274.

[21] Fraiwan, A., D.F. Call, and S. Choi, Bacterial growth and respiration in laminar flow microbial fuel cells. Journal of Renewable and Sustainable Energy, 2014. 6(2): p.-.

[22] Gunawardena, A., S. Fernando, and F. To, Performance of a yeast-mediated biological fuel cell. International journal of molecular sciences, 2008. 9(10): p. 1893-1907.

[23] White, A., P. HAndler, and E.L. Smith, Principles of Biochemistry, ed. 5. 1973: McGraw Hill.

[24] Busalmen, J.P., A. Esteve-Nuñez, and J.M. Feliu, Whole cell electrochemistry of electricity-producing microorganisms evidence an adaptation for optimal exocellular electron transport. Environmental science & technology, 2008. 42(7): p. 2445-2450.

[25] LaBelle, E. and D. Bond, Cyclic voltammetry for the study of microbial electron transfer at electrodes. Bioelectrochemical systems: from extracellular electron transfer to biotechnological application, 2009: p. 137-152.

[26] Kim, H.J., et al., A mediator-less microbial fuel cell using a metal reducing bacterium, Shewanella putrefaciens. Enzyme and Microbial Technology, 2002. 30(2): p. 145-152.

[27] Torres, C.I., et al., A kinetic perspective on extracellular electron transfer by anode-respiring bacteria. FEMS Microbiology Reviews, 2010. 34(1): p. 3-17.

[28] Lovley, D.R., et al., Fe (III) and S0 reduction by Pelobacter carbinolicus. Applied and environmental microbiology, 1995. 61(6): p. 2132-2138.

[29] Prasad, D., et al., Direct electron transfer with yeast cells and construction of a mediatorless microbial fuel cell. Biosensors and Bioelectronics, 2007. 22(11): p. 2604-2610.

[30] Malvankar, N.S., et al., Tunable metallic-like conductivity in microbial nanowire networks. nature nanotechnology, 2011. 6(9): p. 573-579.

[31] Reardon, P.N. and K.T. Mueller, Structure of the Type IVa Major Pilin from the Electrically Conductive Bacterial Nanowires of Geobacter sulfurreducens. Journal of Biological Chemistry, 2013. 288(41): p. 29260-29266.

[32] Reguera, G., et al., Extracellular electron transfer via microbial nanowires. Nature, 2005. 435(7045): p. 1098-1101.

[33] Lies, D.P., et al., Shewanella oneidensis MR-1 uses overlapping pathways for iron reduction at a distance and by direct contact under conditions relevant for biofilms. Applied and Environmental Microbiology, 2005. 71(8): p. 4414-4426.

[34] Gorby, Y.A., et al., Electrically conductive bacterial nanowires produced by Shewanella oneidensis strain MR-1 and other microorganisms. Proceedings of the National Academy of Sciences, 2006. 103(30): p. 11358-11363.

[35] Malvankar, N.S. and D.R. Lovley, Microbial nanowires: a new paradigm for biological electron transfer and bioelectronics. ChemSusChem, 2012. 5(6): p. 1039-1046.

[36] Schröder, U., Anodic electron transfer mechanisms in microbial fuel cells and their energy efficiency. Physical Chemistry Chemical Physics, 2007. 9(21): p. 2619-2629.

[37] Sekoai, P.T. and E.B. Gueguim Kana, Semi-pilot scale production of hydrogen from Organic Fraction of Solid Municipal Waste and electricity generation from process effluents. Biomass and Bioenergy, 2014. 60(0): p. 156-163.

[38] Straub, K.L. and B. Schink, Ferrihydrite-dependent growth of Sulfurospirillum deleyianum through electron transfer via sulfur cycling. Applied and environmental microbiology, 2004. 70(10): p. 5744-5749.

[39] Luo, H., et al., [Isolation and characterization of electrochemical active bacterial Pseudomonas aeruginosa strain RE7]. Huan jing ke xue=Huanjing kexue/ [bian ji, Zhongguo ke xue yuan huan jing ke xue wei yuan hui" Huan jing ke xue" bian ji wei yuan hui.], 2009. 30(7): p. 2118-2123.

[40] Thygesen, A., et al., The effect of different substrates and humic acid on power generation in microbial fuel cell operation. Bioresource Technology, 2009. 100(3): p. 1186-1191.

[41] Yamazaki, S.-i., et al., Role of 2-amino-3-carboxy-1,4-naphthoquinone, a strong growth stimulator for bifidobacteria, as an electron transfer mediator for NAD(P)+regeneration in Bifidobacterium longum. Biochimica et Biophysica Acta (BBA)-General Subjects, 1999. 1428(2–3): p. 241-250.

[42] Liu, H., S. Cheng, and B.E. Logan, Production of Electricity from Acetate or Butyrate Using a Single-Chamber Microbial Fuel Cell. Environmental Science & Technology, 2004. 39(2): p. 658-662.

[43] Rabaey, K., et al., Tubular Microbial Fuel Cells for Efficient Electricity Generation. Environmental Science & Technology, 2005. 39(20): p. 8077-8082.

[44] Larminie, J.D., A., Fuel cell systems explained. 2000: John Wiley & Sons,.

[45] Schröder, U., Niessen, J. and Scholz, F., A generation of microbial fuel cells with current outputs boosted by more than one order of magnitude.. Angewandte Chemienternational Edition. 2003. 42(25): p. 2880-2883.

[46] Rhoads, A., H. Beyenal, and Z. Lewandowski, Microbial Fuel Cell using anaerobic respiration as an anodic reaction and biomineralized manganese as a cathodic reactant. Environmental Science and Technology, 2005. 39: p. 4666-4671.

[47] Aelterman, P., et al., Loading rate and external resistance control the electricity generation of microbial fuel cells with different three-dimensional anodes. Bioresource Technology, 2008. 99(18): p. 8895-8902.

[48] Xie, X., et al., Three-Dimensional Carbon Nanotube−Textile Anode for High-Performance Microbial Fuel Cells. Nano Letters, 2010. 11(1): p. 291-296.

[49] Rozendal, R.A., et al., Towards practical implementation of bioelectrochemical wastewater treatment. Trends in biotechnology, 2008. 26(8): p. 450-459.

[50] Logan, B.E., et al., Microbial fuel cells: methodology and technology. Environmental science & technology, 2006. 40(17): p. 5181-5192.

[51] Zhu, N., et al., Improved performance of membrane free single-chamber air-cathode microbial fuel cells with nitric acid and ethylenediamine surface modified activated carbon fiber felt anodes. Bioresource Technology, 2011. 102(1): p. 422-426.

[52] Zhou, M., et al., Anode modification by electrochemical oxidation: A new practical method to improve the performance of microbial fuel cells. Biochemical Engineering Journal, 2012. 60(0): p. 151-155.

[53] kumar, G.G., V.G.S. Sarathi, and K.S. Nahm, Recent advances and challenges in the anode architecture and their modifications for the applications of microbial fuel cells. Biosensors and Bioelectronics, 2013. 43(0): p. 461-475.

[54] Vargas, I.T., I.U. Albert, and J.M. Regan, Spatial distribution of bacterial communities on volumetric and planar anodes in single-chamber air-cathode microbial fuel cells. Biotechnology and Bioengineering, 2013. 110(11): p. 3059-3062.

[55] Logan, B., et al., Graphite Fiber Brush Anodes for Increased Power Production in Air-Cathode Microbial Fuel Cells. Environmental Science & Technology, 2007. 41(9): p. 3341-3346.

[56] Fan, Y., et al., Nanoparticle decorated anodes for enhanced current generation in microbial electrochemical cells. Biosensors and Bioelectronics, 2011. 26(5): p. 1908-1912.

[57] Xu, S., et al., Enhanced performance and mechanism study of microbial electrolysis cells using Fe nanoparticle-decorated anodes. Applied Microbiology and Biotechnology, 2012. 93(2): p. 871-880.

[58] Feng, C., et al., A dual-chamber microbial fuel cell with conductive film-modified anode and cathode and its application for the neutral electro-Fenton process. Electrochimica Acta, 2010. 55(6): p. 2048-2054.

[59] Luckarift, H.R., et al., Facile Fabrication of Scalable, Hierarchically Structured Polymer/Carbon Architectures for Bioelectrodes. ACS Applied Materials & Interfaces, 2012. 4(4): p. 2082-2087.

[60] Zhao, Y., et al., Hydroxylated and aminated polyaniline nanowire networks for improving anode performance in microbial fuel cells. Journal of Bioscience and Bioengineering, 2011. 112(1): p. 63-66.

[61] Yuan, Y. and S. Kim, Improved performance of a microbial fuel cell with polypyrrole/carbon black composite coated carbon paper anodes. Bull. Korean Chem. Soc, 2008. 29(7): p. 1345.

[62] Higgins, S.R., et al., Fabrication of macroporous chitosan scaffolds doped with carbon nanotubes and their characterization in microbial fuel cell operation. Enzyme and Microbial Technology, 2011. 48(6–7): p. 458-465.

[63] Kalathil, S., et al., Enhanced Performance of a Microbial Fuel Cell Using CNT/MnO2 Nanocomposites as a Bioanode Materials. Journal of nanoscience and nanotechnology, 2013. 13: p. 7712-7716.

[64] Liang, P., et al., Carbon nanotube powders as electrode modifier to enhance the activity of anodic biofilm in microbial fuel cells. Biosensors and Bioelectronics, 2011. 26(6): p. 3000-3004.

[65] Mink, J.E., et al., Vertically Grown Multiwalled Carbon Nanotube Anode and Nickel Silicide Integrated High Performance Microsized (1.25 μL) Microbial Fuel Cell. Nano Letters, 2012. 12(2): p. 791-795.

[66] Yong, Y.-C., et al., Macroporous and Monolithic Anode Based on Polyaniline Hybridized Three-Dimensional Graphene for High-Performance Microbial Fuel Cells. ACS Nano, 2012. 6(3): p. 2394-2400.

[67] Wang, H., et al., High power density microbial fuel cell with flexible 3D graphene-nickel foam as anode. Nanoscale, 2013. 5(21): p. 10283-10290.

Chapter 3

The Anode Biocatalyst with Simultaneous Transition Metals Pollution Control

Oleksandr Bilyy, Oresta Vasyliv and Svitlana Hnatush

Additional information is available at the end of the chapter

http://dx.doi.org/10.5772/58347

1. Introduction

The need for reducing dependence on fossil fuels and the promoting the use of renewable fuels requires the development of alternative sources such as waste biomass for environmental benefits and alternative global energy supplies [1]. Microbial fuel cells (MFCs) provide new opportunities for the sustainable production of energy from biodegradable and reduced compounds, and thus, have attracted substantial research efforts to develop various devices for generating electricity and removing wastes [1, 2]. The development of processes that can use bacteria to produce electricity represents a highly effective method for bioenergy production as the bacteria are self-replicating, and thus the catalysts of organic matter oxidation are self-sustaining [2]. The substrates used in MFCs range from carbohydrates (e.g. glucose, sucrose, cellulose, starch), volatile fatty acids (e.g. formate, acetate, butyrate), alcohols (e.g. ethanol, methanol), amino acids, proteins and even inorganic components such as sulfides or acid mine drainages [2, 3-9]. The type of substrate fed to a MFC potentially has an impact on the structure and composition of the microbial community. Untill now, no clear image of the effect of the type of substrate on electricity generation by the microbial fuel cells is available.

Analysis of external resistances, electron donor concentrations, cell densities, rates of electron transfer to electrodes at various voltages, and anode potentials can aid in understanding the power production capabilities using microorganisms [10]. A simplified model for the conversion of complex organic fuels to electricity is shown in figure 1 [10].

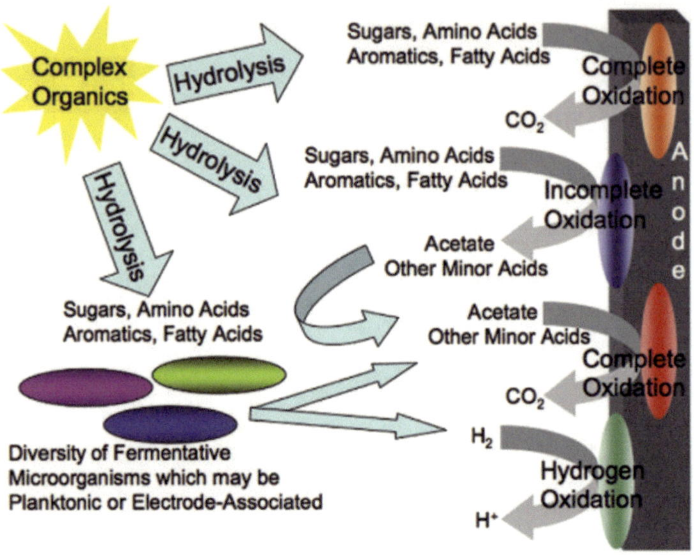

Figure 1. Usage of organic substrates as electron donors in microbial fuel cell [10]

Complex organic matter is hydrolyzed to constituents that in the most cases are primarily fermented, but there are microorganisms that can completely oxidize such compounds with an electrode serving as the sole electron acceptor or incompletely oxidize these substrates with electron transfer to an electrode [10]. Acetate and some other minor fermentation acids can be completely oxidized to carbon dioxide and it is typically the primary source of electrons for current production [10].

MFC is considered to be used also for hydrogen production from the generated potential of the organic matter electrolysis by bacteria [2].

Microbial fuel cell technologies also are a promising and yet completely distinct approach to wastewater treatment as the treatment process can become a method of capturing energy in the form of electricity or hydrogen gas, rather than a drain on electrical energy [2]. Wastewater treatment processes currently employ the biological activities of complex microbial biofilms to remove organic pollutants [11]. The most significant energy savings associated with the use of MFCs for wastewater treatment, besides electricity generation, result from savings in expenses for aeration and solids handling, because the major operating costs for wastewater treatment are wastewater aeration, sludge treatment, and wastewater pumping. The MFC process is inherently an anaerobic process, although, oxygen can diffuse into the system resulting in some aerobic organic matter removal [10].

At the same time, wastewater contains high concentrations of xenobiotics, such as heavy metal ions that have an overwhelming harmful effect towards all living organisms. These substances even in small concentration in the environment cause the increasing inhibition of physiological and biochemical properties of the most bacteria. Despite that, some genera of bacteria possess high resistance according to toxic heavy metals influence because of functioning of highly-efficient mechanisms of antioxidant defense system, ion efflux transport enzyme complexes etc. Examples of these genuses are *Feroplasma, Streptomyces, Thiobacillus, Desulfuromonas* etc [12]. The specific metal-transport systems that were found in gram-negative bacteria which support transport of iron, zinc and manganese, copper, nickel and cobalt are schematically presented in figure 2 [13].

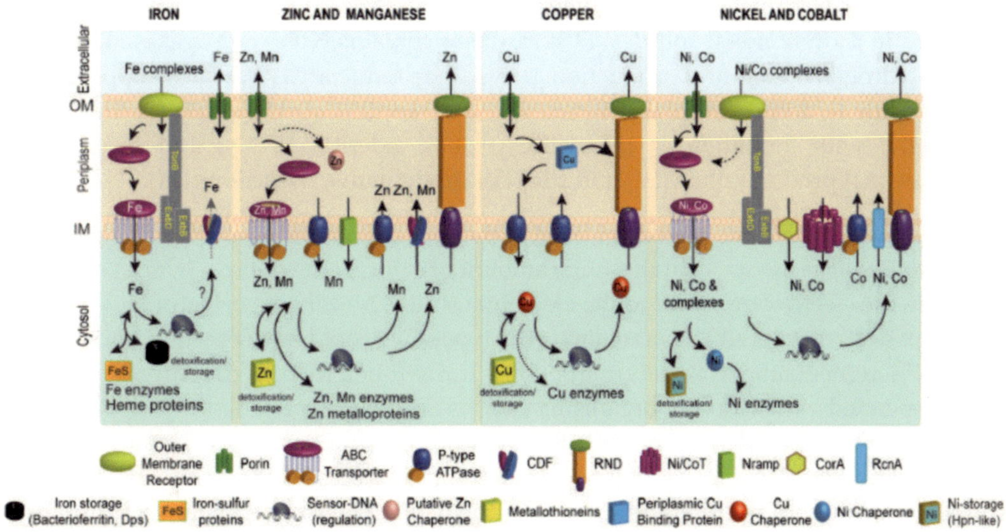

Figure 2. Schematic metal homeostasis models for iron, zinc and manganese, copper, nickel and cobalt in gram-negative bacteria [13]

The variety of ion transport systems in gram-negative bacteria represents sophisticated mechanisms of bacterial cell metal homeostasis regulation. This is possible because of the formation of specific protein-metal coordination complexes used to effect uptake, efflux, intracellular trafficking within compartments, and storage [12].

Previous research has shown that gram-negative bacterium *Desulfuromonas acetoxidans* possesses resistance to copper, cadmium, lead, zinc, manganese, iron, etc. [14, 15]. Bacteria of *Desulfuromonas* genus are also shown to be highly effective in term of electricity generation [2, 16, 17]. However the research on its application as the anode biocatalyst in MFC is inadequate. *Desulfuromonas* sp. can be very promising for MFC development because of inexpensive cultivation medium, high survival rate and resistance to toxic xenobiotics such as the various metal ions [2].

D. acetoxidans, which belongs to the class ∂-*Proteobacteria,* are uncolored obligate anaerobes that inhabit sulfur containing aquatic environments [18]. *D. acetoxidans* supports reductive stage of sulfur cycle in the nature, but it can not possess an ability to reduce sulfate or other oxoanions of this element. Sulfur is a crucial component of various biological active substances, such as vitamins, coenzymes, several amino acids etc. This is an element with variable valence, which participates in the different chemical and biochemical redox-reactions [19]. Sulfur deposits formation is tightly bound with the process of sulfate-reduction, which is carried out exclusively by microorganisms. $SO_4^{2-}$is the prevalent hydrogen acceptor in the processes of organic matter destructions under anaerobic conditions. Majority of strains can use unspecific electron acceptors, such as L-malate or fumarate, instead of sulfur [20]. S^0-reduction by *D. acetoxidans* causes hydrogen sulfide production. Using metal-resistant strains of these bacteria also helps overcome H_2S toxicity since divalent cations will interact with sulfide ions, forming insoluble precipitates in form of metal sulfides. *D. acetoxidans* contains NiFe-hydrogenase [20], which catalyses hydrogen uptake and production; polysulfide reductase [19], which supports sulfur reduction with hydrogen sulphide formation, and specific metaloreductase that reduce $Fe^{3+}$and Mn^{4+} with formation of magnetite (Fe_3O_4), siderite ($FeCO_3$) and rhodochrosite ($MnCO_3$) as final products of Fe (III)-and Mn (IV)-dissimilative reductions [21].

Several redox-proteins have been elucidated of the cells of *D. acetoxidans* [22]. This bacterium, similarly to *Desulfovibrio* sp., contains huge amount of cytochromes *c*-type, which possibly are involved in the electron transport to the elemental sulfur. Multiheme cytochromes *c*-type are shown to possess metaloreductase activity, which possibly could have practical application in metal ions bioremediation from the environment. It was found that electric current production in MFCs operated with different organisms such as *Shewanella oneidensis*, *Pelobacter carbinolicum* or *Geobacter sulfurreducens* requires the presence of multiheme cytochromes *c* [23, 24]. Given the clear interest in *D. acetoxidans* for alternative processes of energy generation, the identification and understanding of the role of the macromolecular components responsible for these metabolic capabilities becomes a priority. Recently the final draft genome of *D. acetoxidans* was made available by the Joint Genome Institute [25] coding for "cytochromome" of 47 putative multiheme cytochromes *c*-type. Of those, up to now only the triheme cytochrome c_7 was characterized in detail. Its structure in the fully oxidized and fully reduced states, its thermodynamic and kinetic redox properties and its thermodynamic stability has been reported in the literature [22]. It shows high similarity to tetraheme cytochrome c_3, extracted from sulfate-reducing bacteria [26]. Cytochrome synthesis by bacteria is observed under usage of insoluble extracellular electron acceptors, such as sulfur.

D. acetoxidans can obtain energy as a result of sulfur respiration and complete acetate oxidation via the citric acid cycle [27]:

$$\text{Acetate}^- + 4S^0 + H^+ + 2H_2O \rightarrow 2CO_2 + 4H_2S;\ \Delta G = -39 kJ \times mol^{-1}$$

This is the first investigated microorganism, which obtains energy by the complete acetate oxidation under the anaerobic conditions. It was shown that only 4% of consumed acetate by bacteria was assimilated into the cell material [28]. *D. acetoxidans* contains a succinyl-CoA: acetate CoA transferase instead of an acetyl-CoA synthetase and a succinyl-CoA synthetase. The succinyl-CoA: acetate-CoA transferase couples the formation of succinate from succinyl-CoA with the activation of acetate. The enzymes required for the assimilation of acetate and CO_2 into pyruvate are acetyl-CoA synthetase and pyruvate synthase [27]. C^{14}-labeling experiments shown that *D. acetoxidans* metabolism includes synthesis of oxaloacetate from acetate and $2CO_2$ as anaplerotic reaction while most organisms oxidizing acetate to CO_2 by using the glyoxylate bypass for this function. Besides acetate, *D. acetoxidans* can completely oxidize L-malate, fumarate, propionate, ethanol etc as electron donors. *D. acetoxidans* was one of the first electrogenic bacterium described, a microorganism performing complete oxidation of an organic substrate with electron transfer directly to the electrode. It was calculated that in the fuel cell that contains acetate as the sole electron donor up to 82% of acetate is oxidized by *D. acetoxidans* with an electrode as the terminal electron acceptor [29]. Therefore, *D. acetoxidans* accumulates energy for growth by electron transfer to the electrodes. Similar results were obtained with *Geobacter metallireducens*, oxidizing aromatic compounds, and the predominantly freshwater *G. sulfurreducens* [17, 30].

Thus, *D. acetoxidans* has a crucial role in the biosphere and shows prospect of prosperous development of microbial fuel cell with simultaneous control of toxic metals environmental pollution because of formation of insoluble precipitates of metal sulfides. Although profound analyses of *D. acetoxidans* physiology and its linkage with power generation in MFC and organic matter consumption need to be established for its efficient application as the anode biocatalyst in microbial fuel cell.

2. Methodology

2.1. Microbial strains, medium and cultivation

Microbial strain *D. acetoxidans* IMV B-7384, which was applied in these investigations, belongs to the Ukrainian Collection of Microorganisms of D.K. Zabolotny Institute of Microbiology and Virology of NAS of Ukraine. Bacteria have been cultivated under the anaerobic conditions in the modified Postgaite C medium [20] in which sterile sulfur (10 g/l) and biotin (20 μg) were added before cultivation. Biotin served as a growth factor. Optimal pH for growth was 6.8-7.5 and optimal temperature was 30 ^{0}C.

2.2. Cell size distribution and relative content measurement

Bacterial growth commonly can be investigated by the registration of bacterial suspension turbidity or by the methods of dynamic or static light dispersion. The new method of rapid measurement of cell size distribution and their relative content, which is based on cell light scattering changes [31, 32] is proposed in this study. It includes the sounding of flow suspended bacterial cells by monochromatic coherent light, the registration of cooperative signals of sounding radiation and the explored microbiological objects by detecting of amplitudes and duration of scattered light impulses. The distribution of particles in size is determined on the basis of the measured functional dependence of the number of registered particles upon the amplitude and duration of corresponding electric impulses on the photoreceiver output by solving integral equation of Fredholm of the first kind (1):

$$F(U,t) = \int_{r_{min}}^{r_{max}} K(U,t,r) n(r) d(r), \quad (1)$$

where r_{min} and r_{max} – upper and lower limits of particle size distribution, which is registered;

n(r) – the function of particle size distribution;

K(U,t,r) – the function of distribution of normalized values of amplitude and duration of registered impulses of scattered light by the calibrating particles, which is a result of the previous probing of liquid flow by the monochromatic coherent light of the polymeric latex with set sizes and known refractive index.

However, presence of bacterial metabolism products in the growth medium could lead to errors in cell size distribution measurement because of additional light scattering. These errors were eliminated in the next way. Bacteria were cultivated in the liquid growth medium. Dependence between quantities of microbial cells and background particles in the growth medium had been determined during the time of bacterial cultivation. Liquid growth medium with and without bacterial cells was diluted in the same proportions by using highly purified liquid (deionized water). Then were registered separately the total quantity of cells and background particles in the highly purified liquid, which contains cells, and the total size distribution of background particles in the growth medium without cells. Then, relative content of bacterial cells was determined in the chosen interval of sizes, which equaled 0.3-1.9 μm. The cells relative content was measured by the calculation of quantity ratio of the set size cells to their total quantity. Specimens for determination of cell size distribution were prepared by dilution of 1 ml of bacterial suspension in 100 ml of deionized water [31, 32]. Measurements have been carried out by using the equipment PRM-6M, which was constructed at the Laboratory of Optical-Electronic Device of Faculty of Electronics of Ivan Franko National University of Lviv. The errors of cell size distribution measurement of constructed equipment are 5%.

2.3. Electron microscopy

After the third day of *D. acetoxidans* IMV B-7384 growth, the cells were harvested by centrifugation (2500 g, 30 min, 4°C) and washed three times in a buffer (50 mM potassium-phosphate buffer, pH 7.5). Intact cells were fixed at 1.5% $KMnO_4$ solution during 20 min under the room temperature (20 °C). Post-fixation was carried out with 1% OsO_4 in cacodylic buffer during 90 min under 0 °C. Fixed cells were washed and dehydrated in solutions with gradient concentrations of ethanol and propylene oxide. Specimens were fixed in epoxy Epon 812. Ultrathin cross-sections were obtained with the ultramicrotome UMTP-6 and contrasted by lead citrate [33]. Electron photographs were obtained by transmission electron microscopes UEMB-100B and PEM-100 at acceleration voltage 75 kV. Final photographs magnification – 10000 times.

2.4. Measurement of catalase, superoxide dismutase activity and intracellular reduced glutathione content

Antioxidant defense system activity has been measured in the cell-free extract after the second, third and fourth day of bacterial growth. Cells were washed by 0.9% NaCl solution and disintegrated on the ultrasonic homogenizer at 22 kHz at 0ºC during five minutes. Cell debris were sedimented by centrifugation at 5640-8800g at 4 ºC during 30 minutes. Catalase activity was measured spectrophotometrically at λ=410 nm by the degree of breakdown of hydrogen peroxide in the cell-free extract [34]. Superoxide dismutase activity was measured by the level of inhibition of 2,3,5-triphenyltetrazolium chloride reduction that follows formazan formation (absorbance maximum λ=405 nm) [35, 36]. Reduced glutathione content has been measured by the degree of dithionitrobenzoic acid reduction in cell-free extract (absorbance maximum λ=412 nm) [37].

2.5. Microbial fuel cell construction and maintenance

In this study two chamber microbial fuel cell has been constructed, in which *D. acetoxidans* IMV B-7384 was applied as the anode biocatalyst. Bacteria were cultivated in the modified Postgaite C medium [20] without sulfates under the anaerobic conditions and temperature 25-28 ^{0}C during twenty days in MFC. 0.1% potassium permanganate solution served as catholyte and bacterial suspension with 0.30±0.05 g/l dry cell weight/liter initial biomass served as anolyte respectively. 0.1% KMnO4 was replaced after 14 days of bacterial growth in MFC. Anode and cathode chambers with 0.3 l volume were separated by proton-exchange membrane (Millipore) with surface area of 2.5 cm^2. Graphite rods with the surface area of 130 cm^2 were applied as electrodes. Graphite is electrically conductive and conforms to the requirements of electrodes in MFC: non-corrosive, highly conductive, large surface area, high porosity etc [2]. Bacteria were cultivated periodically in the anode chamber of constructed MFC under separate addition of such electron donors as acetic, lactic and fumaric acids in form of sodium salts in concentrations 6 and 9 g/l. Electrode material served as the sole electron acceptor. All experiments were conducted under strictly sterile conditions.

2.6. Power output measurements of the microbial fuel cell

Electric current and voltage generation in constructed MFC were determined from the measured voltage drop across the resistor by multimeter DT-830C. The external load resistor with value 2.2 kΩ, which was shown to be the most optimal in constructed microbial fuel cell, was applied. The power output of an MFC was calculated from the measured voltage, E_{MFC}, across the load and the current as (2):

$$P = I \times E_{MFC} \tag{2}$$

Power generation by *D. acetoxidans* IMV B-7384 was investigated during twenty days of bacterial growth under the application of lactic, acetic and fumaric acids in form of their sodium salts.

3. Results and discussion

3.1. Particularities of *D. acetoxidans* IMV B-7384 growth physiology

Biomass accumulation of *D. acetoxidans* IMV B-7384, its cell size distribution and relative content changes have been carried out [14, 15]. Investigated bacterium has been shown to accumulate biomass the most intensively during third-fourth days of its growth. (fig. 3). Since bacterial growth processes were shown to be maximal during third-fourth days of cultivation, possibly, the highest power output in microbial fuel cell should be observed during this period.

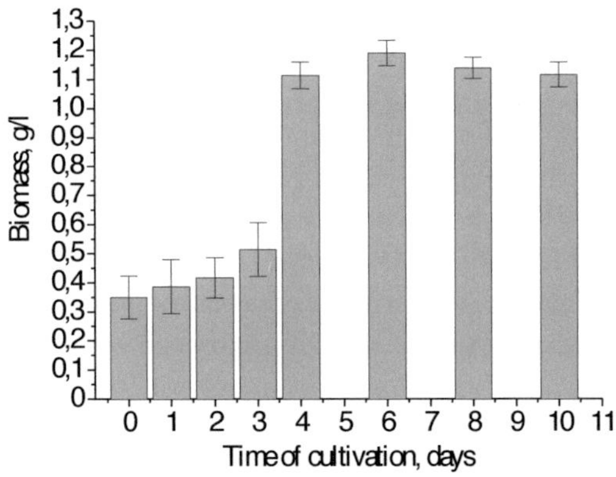

Figure 3. Changes of *D. acetoxidans* IMV B-7384 biomass during ten days under normal cultivation conditions (addition of lactic acid (6 g/l) and elemental sulfur (10 g/l) respectively as electron donor and acceptor)

Electron micrographs of *D. acetoxidans* IMV B-7384 cross-sections are presented on the fig. 4. Cells were obtained on the third day of their growth when maximal bacterial biomass accumulation was observed. These are rod shaped or slightly curved cells.

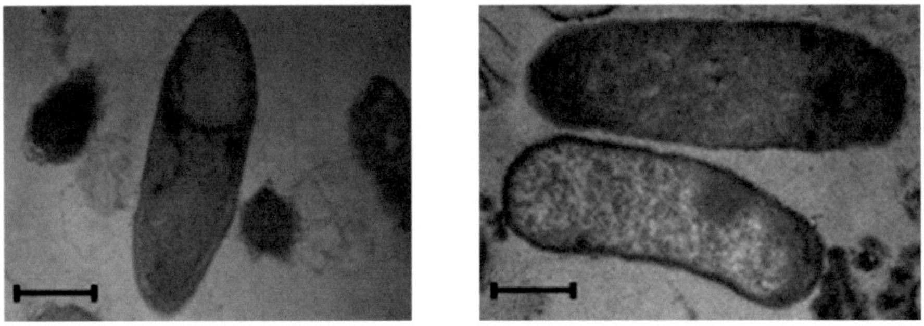

Figure 4. Cross-sections of the cells of *D. acetoxidans* IMV B-7384, harvested under normal growth conditions (third day) (electron micrograph (TEM), ×10000 times; bar, 0.5 μm)

Bacterial growth usually is characterized by increasing of cell quantity or cell size. Thus, analyses of cell size distribution and their relative content allow to obtain more detailed data of cell growth and division processes under various cultivation conditions in comparison with standard turbidimetric method of biomass measurement. Proposed method of cell light scattering determination allows to calculate possible changes of cell division on the basis of cumulative analyses of three histogram parameters: cell size distribution maximum, cell relative content and half-width of cell size distribution. It can be the basis for inventing of new methodologies for obtaining of synchronous cell cultures and also for development of new effective cytometers with low self-cost and high productivity.

Cell size distribution maximum equaled 0.55±0.01 μm on the third day of *D. acetoxidans* IMV B-7384 growth (period of maximal biomass accumulation). Cell relative content increased from 0.275 to 0.420 relative units under normal growth conditions on the third day of bacterial cultivation (fig. 5).

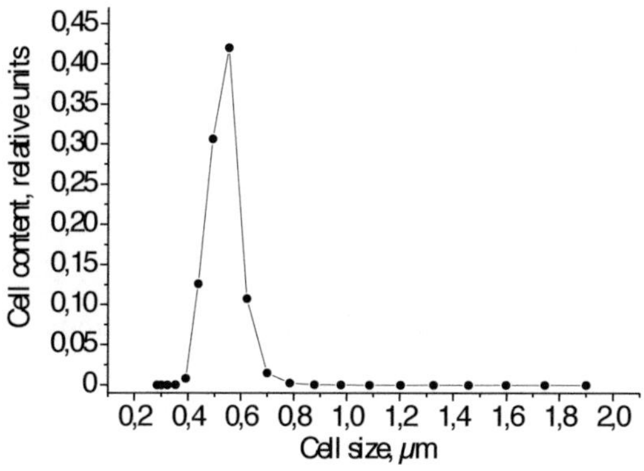

Figure 5. Medial values of *D. acetoxidans* IMV B-7384 cell relative content and size distribution during the third day of growth (five repeats, $P \leq 0.05$)

From the first to the fifth day of *D. acetoxidans* IMV B-7384 growth the maximum of cell size distribution was 0.49-0.55±0.01 μm and cell relative content with the maximum of size distribution changed from 0.275±0.011 to 0.398±0.011 relative units (table 1). Half-width of cell size distribution curves decreased from 0.23±0.01 to 0.14±0.01 μm with the increase of bacterial cultivation time from the first to the third day.

It indicated the decrease of cell size variations with the increase of cultivation time. Obviously, it is caused by intensive bacterial division on the third-fourth days of their cultivation. As a result cell's relative content with lower size distribution maximum (0.49±0.01 μm) increased in comparison with its initial value with higher maximum (0.55±0.01 μm). Possible inhibition of cell division on the fifth day of bacterial growth caused increase of cell relative content with higher size distribution maximum, which equaled 0.55±0.02 μm.

Time of cultivation, day	Cell size distribution maximum, µm	Cell content with size distribution maximum, relative units	Half-width of cell size distribution curves, µm
1	0.55±0.01	0.275±0.011	0.23±0.01
2	0.55±0.02	0.268±0.009	0.23±0.03
3	0.55±0.01	0.420±0.022	0.14±0.01
4	0.49±0.01	0.383±0.14	0.14±0.02
5	0.55±0.02	0.398±0.011	0.16±0.01

Table 1. Changes of cell size distribution maximum, cell relative content with size distribution maximum and half-width of curves of cell size distribution of *D. acetoxidans* IMV B-7384 during five days of cultivation (five repeats, $P \leq 0.05$)

Thus, it was shown that the maximal biomass accumulation is observed during third-fourth day of bacterial growth. High value of cell relative content with size distribution maximum and intensive decrease of half-width of cell size distribution indicates on significant increase of cell quantity with lower size distribution maximum. It is a possible result of intensive cell division during this time. Overall analysis of mentioned above parameters allow to assume that *D. acetoxidans* IMV B-7384 are in the middle of exponential phase of growth during the third day of their cultivation under normal growth conditions. Thus, the most intensive biosynthesis processes and power generation in microbial fuel cell possibly overlap this period of bacterial cultivation or immediately afterwards.

3.2. Response of *D. acetoxidans* IMV B-7384 cells to the influence of heavy metals

The activity of antioxidant defense system of *D. acetoxidans* IMV B-7384 has been revealed under the influence of external stress factors, such as Ferric iron, Ferrous iron, Nickel, Cobalt and Copper salts. It includes:

- biosynthesis of reduced glutathione (GSH), a tripeptide that serves as universal electron donor detoxifying reactive oxygen species [38];
- activity of catalase and superoxide dismutase, a basic enzymes of antioxidant defense system [36, 39].

Reactive oxygen species (ROS) such as superoxide radical, hydrogen peroxide, hydroxyl radical etc are produced as a result of prolonged influence of oxygen or toxic xenobiotics, such as heavy metal ions, on the living cell. Glutathione (γ-L-glutamyl-L-cysteinylglycine) is the most abundant intracellular thiol-dependent antioxidant, which protects living cells against oxidative stress. It has low redox-potential (E'_0=-240 mV under pH=7.0) and it is constantly maintained in reduced state because of NADF-glutathione reductase functioning. Therefore it serves as cellular redox-buffer [40]. Superoxide dismutase catalyzes disproportionation of superoxides with oxygen and hydrogen peroxide formation. Catalase causes decomposition of produced H_2O_2 with neutralization of its toxicity. H_2O and O_2 are final products of this reaction [41]. Thus, aforementioned components of antioxidant defense system protect the cell against oxidative stress, which may be caused by reactive oxygen species.

It was revealed that under the influence of various concentrations of $FeSO_4$, $FeCl_3$, $MnCl_2$, $NiCl_2$, $CoCl_2$ and $CuCl_2$ on the cells of *D. acetoxidans* IMV B-7384 activity of antioxidant defense system appears in spite of their obligate anaerobic metabolism. At fig. 6 is presented the complete scheme of the dependences between the highest values of catalase and superoxide dismutase activity and glutathione content, which were observed in the cells of investigated bacteria during four days of their cultivation.

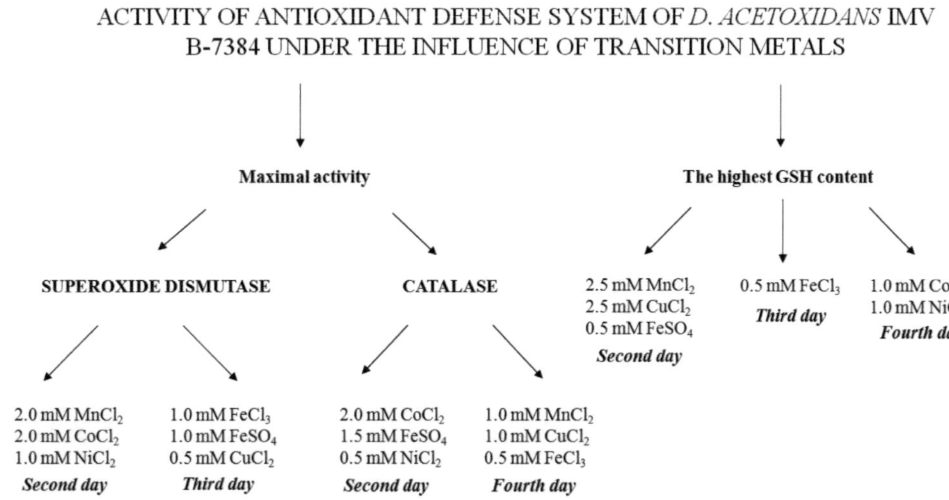

Figure 6. Maximal observed values of reduced glutathione content and superoxide dismutase, and catalase activity in the cell-free extract of *D. acetoxidans* IMV B-7384 under the influence of 0.5-2.5 mM of $FeSO_4$, $FeCl_3$, $MnCl_2$, $NiCl_2$, $CoCl_2$ and $CuCl_2$ during four days of bacterial cultivation [36, 38, 39]

Obtained results show that investigated bacteria possess specific mechanisms of rapid defense against toxic influence of external factors, such as high concentrations of various metal ions. It allows to assume their tolerance according to detrimental xenobiotics, which wastewaters are enriched with. This shows the prospect of *D. acetoxidans* IMV B-7384 application into wastewater treatment with simultaneous power generation in MFC. Also specific activity of antioxidant defense system enzymes, such as catalase and superoxide dismutase could help to prevent harmful influence of reactive oxygen species on the integrity of proton-exchange membranes in microbial fuel cells. Since catalase causes two-electron decomposition of H_2O_2 with O_2 and H_2O production, this enzyme could prevent Fenton reaction: $M^{n+}(=Cu^+, Fe^{2+}, Ti^{3+}, Co^{2+})+H_2O_2 \rightarrow M^{(n+1)+}(=Cu^+, Fe^{2+}, Ti^{3+}, Co^{2+})+{}^{\bullet}OH+OH^-$[42]. Neutralization of products of this reaction could significantly increase longevity of expensive proton-exchange membranes in microbial fuel cell, such as Nafion, with the next increasing of reliability and durability of power generation in MFC.

3.3. Power generation by *D. acetoxidans* IMV B-7384 in constructed microbial-anode fuel cell

3.3.1. The influence of external load resistance on power generation by D. acetoxidans IMV B-7384 as the anode biocatalyst in constructed MFC

External load resistance in microbial fuel cell is one of the crucial factors, which influence the electricity generation. Correlation between value of external load resistance and power generation in constructed MFC was determined. The influence of changes of external load resistance on volt-ampere characteristic of constructed MFC was investigated during *D. acetoxidans* IMV B-7384 cultivation in Postgaite C medium under normal growth conditions on the third-beginning of the fourth day, when the highest power generation was observed (fig. 7).

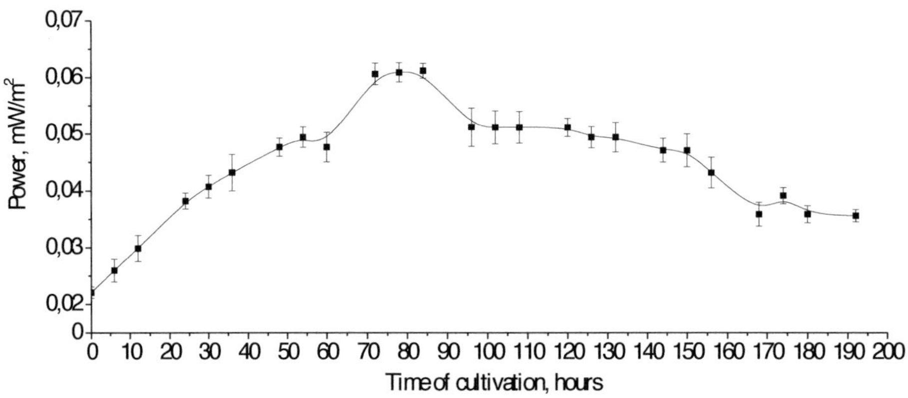

Figure 7. Power generation by *D. acetoxidans* IMV B-7384 during eight days under normal cultivation conditions in MFC, and application of 0.2 kΩ external load resistor

The maximal value of power density equaled 4.7 mW/m^2 on 84 hour of bacterial cultivation with addition of lactate (6 g/l) as electron donor and elemental sulfur (10 g/l) as electron acceptor (normal growth conditions), and application of an external load resistor of 0.2 kΩ. With the increase of cultivation time till 192 hour (eighth day of growth) generated power decreased by 42% in comparison with its maximal value.

With the aim to determine the most optimal load in term of electricity generation by *D. acetoxidans* IMV B-7384 in constructed MFC, external load resistors with values of 0.45-8.15 kΩ were applied in this investigation. The highest current strength and voltage were observed in constructed MFC under application of external load resistor with value of 2.2 kΩ (fig. 8). The highest power density, which was obtained in MFC under these conditions, equaled 5.8 mW/m^2. Increasing of external load resistor value caused decrease of generated power.

Therefore, all further experiments on MFC development were carried out with application of such external load resistance (2.2 kΩ), which was shown to be the most effective in term of electric power generation by *D. acetoxidans* IMV B-7384 in constructed MFC.

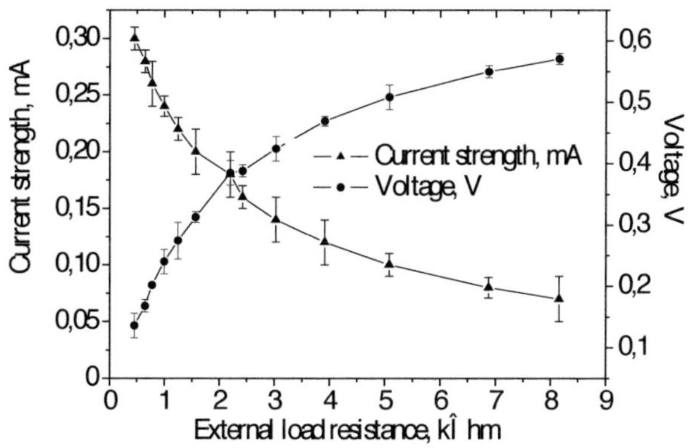

Figure 8. Volt-ampere characteristic of constructed MFC under the influence of various external load resistances on the third-beginning of the fourth days of *D. acetoxidans* IMV B-7384 growth under normal cultivation conditions

Accurate selection of external load resistance plays an important role in effective MFC development, because it significantly influences the value of generated power by bacteria, which are cultivated under different conditions in constructed MFC.

3.3.2. Power output in MFC under usage of various electron donors by D. acetoxidans IMV B-7384 in MFC

3.3.2.1. Lactic acid and power output in MFC

In previous researches it was shown that maximal power output equaled 4.3 mW/m^2 under addition the external electron acceptors, such as sulfur and Ferric iron in concentrations, which are favorable for *D. acetoxidans* IMV B-7384 metabolism [43]. Excluding of these additional electron acceptors and external load resistance optimization caused increasing of power output in constructed MFC.

Lactic acid was applied as the sole organic electron donor in concentrations 6 and 9 g/l during *D. acetoxidans* IMV B-7384 cultivation in the anode chamber of constructed microbial fuel cell. The highest power output equaled 5.74±0.29 mW/m^2 on the 64 hour (third day) of *D. acetoxidans* IMV B-7384 cultivation under addition of 6 g/l of $C_3H_6O_3$ (fig. 9). Its value decreased by 41% in comparison with the maximal power, obtained in this investigation by 250 hour of bacterial cultivation (tenth day). Power output equaled 1.33±0.18 mW/m^2 on the 480 hour of bacterial cultivation (twentieth day), which was lower by 77% in comparison with its maximal value.

Thus, gradual lactate oxidation and the following diminishing of its quantity in the anode chamber because of bacterial growth in MFC caused gradual decrease of produced power with the increase of duration of cultivation time. Thus, application of lactic acid as the single electron donor in the aforementioned concentration in constructed MFC can't be used for sustainable long-term electricity generation.

Concentration of lactic acid in the anode chamber has been increased up to 9 g/l. Under these cultivation conditions the highest power output equaled 5.90±0.21 mW/m^2 on the 64 hour (third day) of *D. acetoxidans* IMV B-7384 cultivation (fig. 10). After ten days it decreased by 37% in comparison with its highest measured value. By the 480 hour (the twentieth day) of *D. acetoxidans* IMV B-7384 cultivation power production in constructed MFC decreased by 33% in comparison with its maximal measured value.

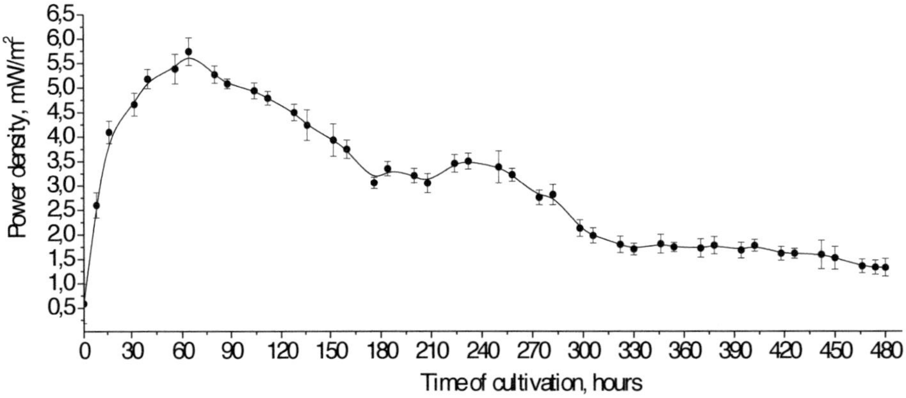

Figure 9. Power density in MFC during twenty days under addition of 6 g/l of lactic acid into the growth medium of *D. acetoxidans* IMV B-7384

Increase of lactic acid content in the anode chamber caused enhance of stability of power generation in constructed MFC, but such manipulation did not boost its value significantly in comparison with application of lower concentration of lactic acid.

Thus, lactic acid as the sole electron donor in high concentrations supports durability of constructed MFC with application of *D. acetoxidans* IMV B-7384 as the anode biocatalyst. However, its effectiveness does not change significantly under the application of lower and higher lactic acid concentrations.

3.3.2.2. Fumaric acid and power output in MFC

With the aim to determine the most optimal electron donor in term of electricity generation lactic acid has been substituted by fumaric acid in the anode chamber of constructed MFC. The highest value of generated power equaled 5.69±0.29 mW/m^2 on the 56 hour (the third day) of bacterial cultivation under usage of 6 g/l of fumaric acid as the sole electron donor (fig. 11). It's

http://dx.doi.org/10.5772/58347

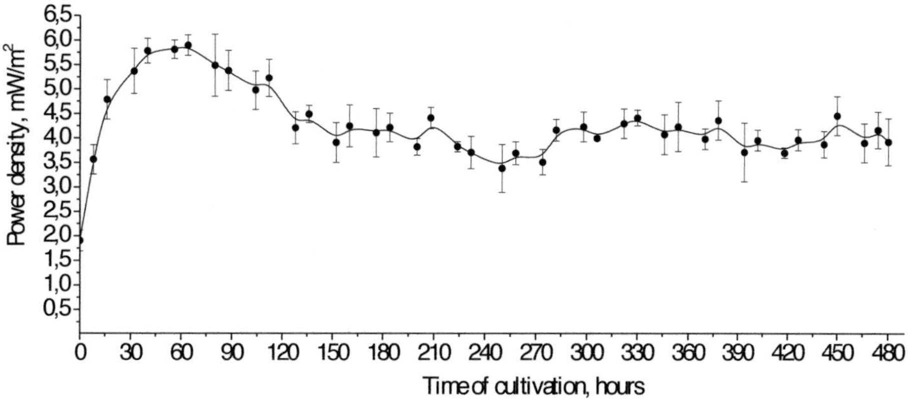

Figure 10. Power density in MFC during twenty days of *D. acetoxidans* IMV B-7384 cultivation under application of 9 g/l of lactic acid

value decreased by 38% on the 250 hour (the tenth day) and by 43% on the 480 hour (the twentieth day) of *D. acetoxidans* IMV B-7384 cultivation. The medial power density, which was observed under these cultivation conditions equaled 3.58±0.20 mW/m^2.

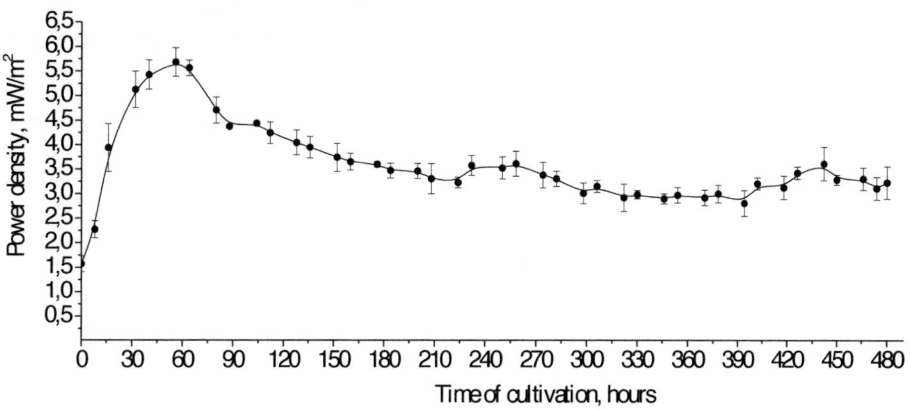

Figure 11. Power density in MFC during twenty days under addition of 6 g/l of fumaric acid into the growth medium of *D. acetoxidans* IMV B-7384

Thus, application of fumaric acid (6 g/l) causes higher stability of power generation by investigated bacteria in constructed MFC in comparison with the lactic acid in the same concentration. Therefore, fumaric acid is more preferable electron donor in term of power generation by *D. acetoxidans* IMV B-7384 in comparison with application of lactic acid.

It was shown that increase of fumaric acid concentration up to 9 g/l caused maximal power generation (2.38±0.10 mW/m^2) at 56 hour (fig. 12).

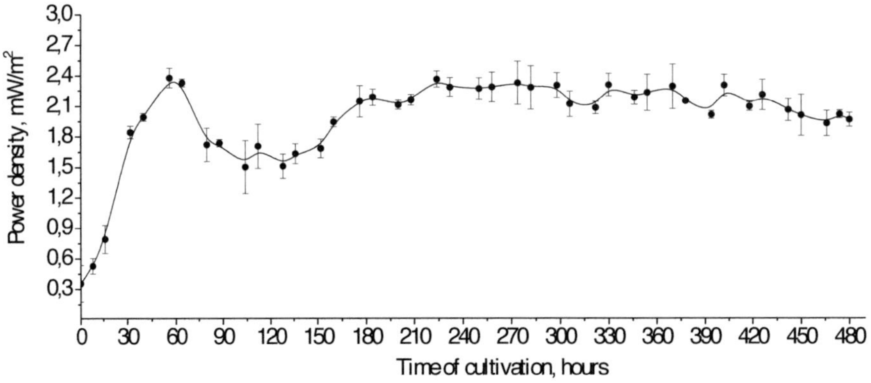

Figure 12. Power density in MFC during twenty days under addition of 9 g/l of fumaric acid into the growth medium of *D. acetoxidans* IMV B-7384

It was less by 2.4 times in comparison with the highest power density value, which was observed under bacterial cultivation with application of less concentration of fumaric acid (6 g/l). The minimal observed power density value equaled 1.95±0.047 mW/m^2 on the 160 hour of bacterial cultivation under these conditions. Increasing of cultivation time caused insignificant enhance of power production. On the 480 hour it value equaled 1.97±0.07 mW/m^2, which was less by 17% in comparison with its maximal measured value in this investigation.

Thus, increase of fumaric acid concentration from 6 to 9 g/l in the anode chamber of constructed MFC reduced its productivity but enhanced its stability in comparison to the application of lower concentration of investigated electron donor.

3.3.2.3. Acetic acid and power output in MFC

Acetic acid was applied as the separate electron donor in constructed MFC. It was shown that the maximal power value equaled 5.78±0.24 mW/m^2 on the 40 hour (second day) of bacterial cultivation under addition of 6 g/l of CH_3COOH (fig. 13).

Its value decreased by 42% on the 232 hour (10 day) of *D. acetoxidans* IMV B-7384 growth. Generated power insignificantly enhanced with the increase of cultivation time. It equaled 3.1±0.12 mW/m^2 on the 480 hour of bacterial cultivation. Thus, application of acetic acid as well as fumaric acid in low concentrations caused high stability of electricity generation in constructed fuel cell apart from application of lactic acid.

D. acetoxidans IMV B-7384 has been cultivated in MFC under addition of 9 g/l of acetic acid as the sole electron donor into the growth medium (fig. 14). Under these cultivation conditions the maximal power value equaled 3.6±0.30 mW/m^2 on the 64 hour (third day) of bacterial cultivation. It was lower by 61% in comparison with the maximal power value obtained under

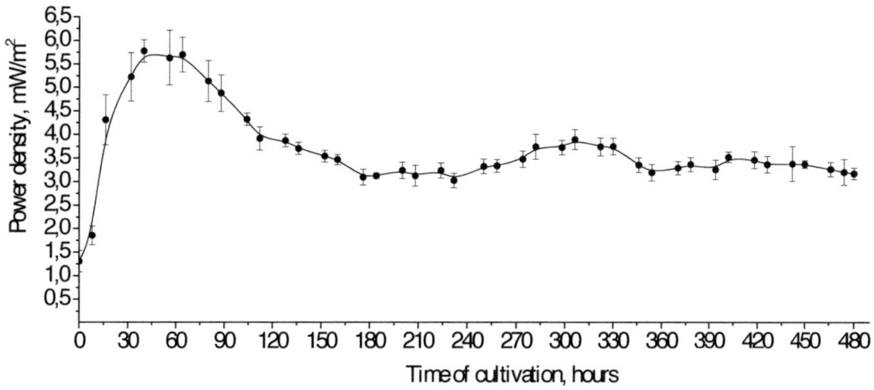

Figure 13. Power density in MFC during twenty days under addition of 6 g/l of acetic acid into the growth medium of *D. acetoxidans* IMV B-7384

usage of 6 g/l of lactic acid by bacteria in the anode chamber of MFC. Increase of cultivation time up to 480 hour caused decrease of power value till 1.60±0.11 mW/m^2.

Thus, increase of acetic acid concentration as electron donor in the anode chamber of MFC caused partial inhibition of electricity generation in comparison with application of its lower content.

It can be summarized that low concentrations of investigated organic acids caused higher stability of power generation in constructed microbial fuel cell apart from their higher concentrations.

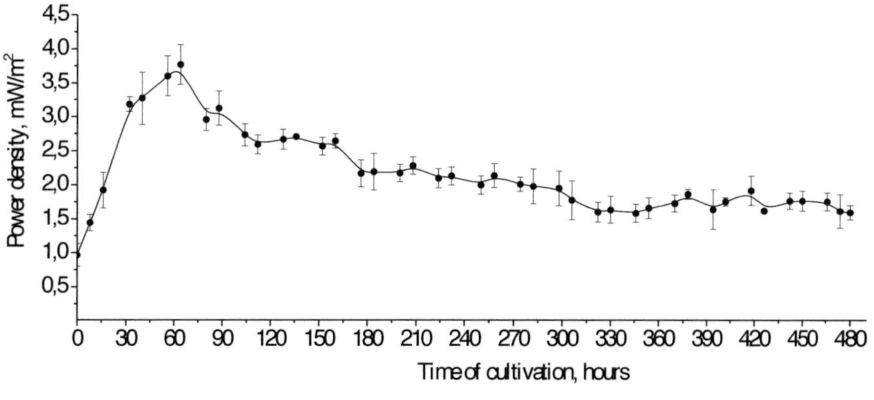

Figure 14. Power density in MFC during twenty days under addition of 9 g/l of acetic acid into the growth medium of *D. acetoxidans* IMV B-7384

Possibly, it may be explained because of raising of by-products concentrations in the growth medium under increasing of organic source concentration. It may cause negative influence according to *D. acetoxidans* IMV B-7384 metabolism and their respective ability of exoelectrogenesis.

4. Conclusions

D. acetoxidans IMV B-7384 is exoelectricigenic sulfur-reducing bacterium which influences environmental biogeochemistry by maintenance of reductive stage of sulfur cycle. Its metal-resistant strains play significant role in heavy metal ions remediation from the aquatic environments because of interaction between the final product of bacterial dissimilative sulfur-reduction – hydrogen sulfide and metal ions with their next combining in form of insoluble metal sulfide precipitates. It was shown that *D. acetoxidans* IMV B-7384 synthesizes such components of antioxidant defense system as catalase, superoxide dismutase and reduced glutathione under the influence of aggressive external factors, such as heavy metal ions. It causes its resistance against environmental pollution by these xenobiotics. Enzymatic and non-enzymatic components of antioxidant defense system found in the cells of *D. acetoxidans* IMV B-7384 are highly effective in neutralization of reactive oxygen species. Antioxidant system activity of investigated bacterial cells may be useful for increasing the durability of proton-exchange membranes in MFCs because of the creation of defensive barrier against detrimental influence of these oxidants. It shows a prospect of efficient and economic application of *D. acetoxidans* IMV B-7384 as the anode biocatalyst in microbial fuel cell with simultaneous wastewater treatment.

The optimal external resistance in constructed MFC in term of power generation was determined to be 2.2 kΩ. Separate application of lactic, fumaric, and acetic acids caused differences in power generation by investigated bacterium. It was shown that addition of fumaric and acetic acids in concentration 6 g/l improved stability of generated power in constructed microbial fuel cell in comparison with application of lactic acid in the same concentration. Increase of concentration of investigated organic electron donors up to 9 g/l reduced generated electric power .

Thus, *D. acetoxidans* IMV B-7384 may be applied for effective treatment of wastewater enriched with heavy metals, acetic and fumaric acids-containing refuses with simultaneous electricity generation in the scaled-up microbial fuel cells. Additionally it can be used for treatment of highly polluted sulfur-containing aquatic environments with alterations of sulfur cycle.

4.1. Possible directions of further research on application of *D. acetoxidans* IMV B-7384 as the anode biocatalyst in MFC

Exploration of the utility of *D. acetoxidans* IMV B-7384 for development of efficient MFC through determination of optimal cultivation and fuel cell construction parameters may be highly beneficial for the progress of microbial fuel cell study with simultaneous heavy metals pollution control for cost effective environmental remediation. Further research of *D. acetoxi-*

dans IMV B-7384 as the anode biocatalyst may include detailed analyses of its molecular biochemistry with the aim of profound understanding of interconnections between the processes of organic source consumption and electric current generation. Further analyses of interrelations between specific reactions of sulfur cycle (e.g. polysulfide reductase activity), reduction of transition metals, such as iron and manganese, and processes of electrogenesis, which are conducted by the cells of *D. acetoxidans* IMV B-7384 may substantially influence the microbial fuel cell study with the aim of increasing of its productivity, reliability and durability.

Acknowledgements

We are highly grateful to Dr. Neelkanth G. Dhere, Jaroslav Ferensovych, Dr. Vasyl Getman, Dr. Dariya Fedorovych, Dr. Yurij Boretsky, Dr. Oleksandr Kulachkovskyj, and all other our coworkers for their support provided in carrying out of investigations and book chapter preparation.

Author details

Oleksandr Bilyy[1*], Oresta Vasyliv[2] and Svitlana Hnatush[2]

*Address all correspondence to: bily2011@yandex.ru

1 Ivan Franko National University of Lviv, Faculty of Electronics, Laboratory of Optical-Electronic Device, Ukraine

2 Ivan Franko National University of Lviv, Biological Faculty, Department of Microbiology, Ukraine

References

[1] Zhuwei D., Haoran L., Tingyue G. A state of the art review on microbial fuel cells: A promising technology for wastewater treatment and bioenergy. Biotechnology Advances 2007; 25, 464-482.

[2] Logan BE. Microbial Fuel Cells. John Wiley & Sons., New Jersey; 2007.

[3] Cheng S., Dempsey BA., Logan BE. Electricity generation from synthetic acid-mine drainage (AMD) water using fuel cell technologies. Environmental Science & Technology 2007; 41, 8149-8153.

[4] He Z., Minteer SD., Angenent LT. Electricity generation from artificial wastewater using an upflow microbial fuel cell. Environmental Science & Technology 2005; 39, 5262-5267.

[5] Heilmann J., Logan BE. Production of electricity from proteins using a microbial fuel cell. Water Environment Research 2006; 78, 531-537.

[6] Logan BE., Murano C., Scott K., Gray ND., Head IM. Electricity generation from cysteine in a microbial fuel cell. Water Research 2005; 39, 942-952.

[7] Min B., Logan BE. Continuous electricity generation from domestic wastewater and organic substrates in a flat plate microbial fuel cell. Environmental Science & Technology 2004; 38, 5809-5814.

[8] Rabaey K., Lissens G., Siciliano SD., Verstraete W. A microbial fuel cell capable of converting glucose to electricity at high rate and efficiency. Biotechnology Letters 2003; 25, 1531-1535.

[9] Rabaey K., Clauwaert P., Aelterman P., Verstraete W. Tubular microbial fuel cells for efficient electricity generation. Environmental Science & Technology 2005; 39, 8077-8082.

[10] Lovley DR. The microbe electric: conversion of organic matter to electricity. Current Opinion in Biotechnology 2008; 19, 1-8.

[11] Wineld J., Ieropoulos I., Greenman J., Dennis J. The overshoot phenomenon as a function of internal resistance in microbial fuel cells. Bioelectrochemistry 2011; 81(1), 22-27.

[12] Winkelmann G. Microbial Transport Systems. New York: WILEY–VCH; 2002.

[13] Ma Zh., Jacobsen FE., Giedroc DP. Metal transporters and metal sensors: how coordination chemistry controls bacterial metal homeostasis. Chemical Reviews 2009; 109 (10) 4644-4681.

[14] Vasyliv OM., Bilyy OI., Getman VB., Ferensovych YP., Yaremyk RY., Hnatush SO. The influence of different metal ions on light scattering properties of pattern microbial fuel cells' bacteria *Desulfuromonas acetoxidans*. in: Dhere NG., Wohlgemuth JH., Lynn KW. (eds.) Optics+Photonics / Solar Energy+Technology: Reliability of Photovoltaic Cells, Modules, Components, and Systems IV: proceedings of the International Symposium of the Society of Photo-Optical Instrumentation Engineers, 21-25 August 2011, SPIE, Bellingham, WA; 2011.

[15] Vasyliv OM., Bilyy OI., Getman VB., Kushkevych IV., Hnatush SO. The changes of spectroscopic characteristics of sulfur-reducing bacteria *Desulfuromonas acetoxidans* under the influence of different metal ions. in: Hoover RB., Davies PCW., Levin GV. (eds.) Optics+Photonics / Optical Engineering+Applications: Instruments, Methods, and Missions for Astrobiology XIV: proceedings of the International Symposium of

the Society of Photo-Optical Instrumentation Engineers, 21-25 August 2011, SPIE, Bellingham, WA; 2011.

[16] Tender LM., Gray SA., Groveman EG. The first demonstration of a microbial fuel cell as a viable power supply: Powering a meteorological buoy. Journal of Power Sources 2008; 179 (3) 571-575.

[17] Debabov VG. Electricity from Microorganisms. Microbiology 2008; 77 (2) 123-131.

[18] Garity G., Winters M., Searles D. Taxonomic outline of the procariotic genera Bergey's manual of systematic bacteriology. Second edition. Michigan. USA: Springer-Verlag; 2001.

[19] Peck JrHD., Lissolo T., Cole JA., Ferguson SJ. The Nitrogen and Sulphur Cycles. Cambridge: Cambridge University Press; 1988.

[20] Brugna M., Nitschke W., Toci R., Bruschi M., Giudici-Orticoni M.-Th. First evidence for the presence of a hydrogenase in the sulfur-reducing bacterium *Desulfuromonas acetoxidans* // Journal of Bacteriology 1999; 181 (17) 5505-5508.

[21] Roden EE., Lovley DR. Dissimilatory Fe (III) reduction by the marine microorganism *Desulfuromonas acetoxidans.* Applied and Environmental Microbiology 1993; 59 (3) 734-742.

[22] Alves AS., Paquete CM., Fonseca BM., Louro RO. Exploration of the cytochromome of *Desulfuromonas acetoxidans*, a marine bacterium capable of powering microbial fuel cells. Metalomics 2011; 3, 349-353.

[23] Haveman SA., DiDonato RJ., Villanueva L., Shelobolina ES., Postier BL., Xu B., Liuand A., Lovley DR. Genome-wide gene expression patterns and growth requirements suggest that *Pelobacter carbinolicus* reduces Fe(III) indirectly via sulfide production. Applied and Environmental Microbiology 2008; 74, 4277-4284.

[24] Shi L., Squier TC., Zachara JM., Fredrickson JK. Respiration of metal (hydr)oxides by *Shewanella* and *Geobacter*: a key role for multihaem *c*-type cytochromes. Molecular Microbiology 2007; 6512-6520.

[25] DOE Joint Genome Institute. *Desulfuromonas acetoxidans* DSM 684 Home: http://genome.jgi-psf.org/desac/desac.home.html (accessed 24 May 2010)

[26] Aubert C., Lojou E., Bianco P., Rousset M., Durand M.-Cl., Bruschi M., Dolla A. The *Desulfuromonas acetoxidans* triheme cytochrome c_7 produced in *Desulfovibrio desulfuricans* retains its metal reductase activity. Applied and Environmental Microbiology 1998; 64 (4) 1308-1312.

[27] Gebhardt NA, Thauer RK., Linder D., Kaulfers P.-M., Pfennig N. Mechanism of acetate oxidation to CO_2 with elemental sulfur in *Desulfuromonas acetoxidans.* Archives of Microbiology 1985; 141, 392-398.

[28] Lovley RD., Holmes DE., Nevin KP. Dissimilatory Fe (III) and Mn (IV) reduction. Advances in Microbial Physiology 2004; 49, 219-286.

[29] Bond DR., Holmes DE., Tender LM., Lovley DR. Electrode-reducing microorganisms that harvest energy from marine sediments. Science 2002; 295, 483-485.

[30] Bond DR., Lovley DR. Electricity production by *Geobacter sulfurreducens* attached to electrodes. Applied and Environmental Microbiology 2003; 69, 1548-1555.

[31] Bilyy OI., Getman VB., Matvijchuk JM. The method of determination of microparticles distribution under their sizes in the disperse environments. Measurement and calculation techniques in the technological processes 2001; 2, 23-26.

[32] Bilyy OI., Getman VB., Kotsyumbas IJ., Rozko MS. New optical method of the monitoring of bacteria in the liquid solutions. Biophysical visnyk 2001; 1(8), 121-126.

[33] Reynolds ES. The use of lead citrate at high pH as an electronopaque stain in electron microscopy. The Journal of Cell Biology 1963, 17, 208-212.

[34] Goth LA. Simple method for determination of serum catalase activity and revision of reference range. Clinics Chimica Acta 1991; 196, 143-152.

[35] Bernas T., Dobrucki JW. The role of plasma membrane in bioreduction of two tetrazolium salts, MTT, and CTC. Archives of Biochemistry and Biophysics 2000; 380 (1) 108-116.

[36] Vasyliv OM., Hnatush SO. Influence of transition metal compounds on superoxide dismutase activity of sulfur-reducing *Desulfuromonas acetoxidans* bacteria. Mikrobiologichny Zhurnal 2013; 75 (2) 37-44.

[37] Owen JB., Butterfield DA. Measurement of oxidized/reduced glutathione ratio. Methods Mol Biol. 2010; 648, 269-77.

[38] Vasyliv OM., Hnatush SO. Effect of ferrum and manganese compounds on the glutathione content in cells of sulfurreducing bacteria *Desulfuromonas acetoxidans.* Studia Biologica 2011; 5 (1) 5-10.

[39] Vasyliv OM., Hnatush SO. Effect of transition metal compounds on catalase activity of sulfur-reducing bacterial *Desulfuromonas acetoxidans* cells. Visnyk of L'viv University. Biological Series 2011; 57, 207-215.

[40] Cotgreave IA., Gerdes RG. Recent trends in glutathione biochemistry – gluthathione-protein interactions: a molecular link between oxidative stress and cell proliferation? Biochemical and Biophysical Research Communication 1998; 242, 1-9.

[41] Brioukhanov AL., Thauer RK., Netrusov AI. Catalase and superoxide dismutase in the cells of strictly anaerobic microorganisms. Microbiolology 2002; 71 (3) 281-285.

[42] Hongsirikarn K., Mo X., Goodwin Jr. JG., Creager S. Effect of H_2O_2 on Nafion® properties and conductivity at fuel cell conditions. Journal of Power Sources 2001; 196, 3060-3072.

[43] Vasyliv OM., Bilyy OI., Ferensovych YP., Hnatush SO. Electric current generation by sulfur-reducing bacteria in microbial-anode fuel cell. in: Dhere NG., Wohlgemuth JH., Lynn KW. (eds.) Optics+Photonics / Solar Energy+Technology: Reliability of Photovoltaic Cells, Modules, Components, and Systems IV: proceedings of the International Symposium of the Society of Photo-Optical Instrumentation Engineers, 12-26 August 2012, SPIE, Bellingham, WA; 2012.

Chapter 4

Anode Biofilm

Michal Schechter, Alex Schechter,
Shmuel Rozenfeld, Emanuel Efrat and Rivka Cahan

Additional information is available at the end of the chapter

http://dx.doi.org/10.5772/58432

1. Introduction

Microbial fuel cells (MFCs) have long been considered an attractive mean for converting various carbohydrate wastes directly into electricity using electrogenic bacterial cells in the anode compartment. Most MFCs have been operated using anaerobic or facultative aerobic bacteria which oxidize various substrates including glucose, sewage sludge and petroleum hydrocarbon [1]. Power production by MFCs varies with bacterial cell species, specific substrate concentration, cathode catalysts and the MFC configuration [2].

Typically, MFCs which were operated with a mixture of bacterial cells produced higher specific power than MFCs operated by a monoculture in the anode compartment [1]. Knowledge and understanding of the anode biofilm components, morphology formation steps and electron transfer mechanism may lead to better biofilm conductivity in MFC.

This chapter focused on biofilm definition and composition. Spectroscopic methods for anode biofilm study, including advancing real-time analysis. Power-producing bacterial cells, mechanisms of electron transfer from the biofilm to the anode and the effect of medium and pH conditions on power production.

2. Biofilm definition

The term biofilm has been proposed for a structured community of microorganisms that adheres irreversibly to surfaces (biotic or abiotic) and is enclosed in a self-developed polymeric matrix of a primarily polysaccharide material [3].

3. Biofilm composition

The biofilm matrix, which is a prerequisite for biofilm formation, consists of up to 97% water, 2-5% microbial cells, 3-6% extracellular polymeric substance (EPS) and ions [4-6]. The EPS may be hydrophilic or hydrophobic and is composed of 40-95% polysaccharides, 1-60% proteins, 1-10% nucleic acids and 1-40% lipids [7].

The EPS serves as a scaffold which holds the cell aggregates together [4]. EPS is highly hydrated, since it can incorporate large amounts of water molecules by hydrogen bonding. The microbial consortia and the environmental conditions influence the composition of the EPS [8]. The amount and thickness of the EPS increase with the biofilm's age [9].

The EPS in thin biofilms is often rich in proteins, contrary to thicker biofilms [10]. The EPS is more abundant in the interior of the biofilm, whereas cell densities are highest in the top layer [11].

As mentioned, the EPS is comprised of exopolysaccharides, proteins, nucleic acids and lipids. The exopolysaccharides in the EPS can be linear or branched, with a molecular weight of 500-2000 kDa. There are homo-polymers, e.g. cellulose, curdlan, dextran and sialic acid, but the majority are hetero-polymers composed of 2-4 types of mono-sugars such as alginate, emulsan, gellan and xanthan [6]. Nucleic acids that are found in the EPS are extracellular DNA (eDNA) that exhibit some similarities to genomic DNA but also distinct differences. In *Pseudomonas aeruginosa*, the release of eDNA is under the control of quorum sensing systems. The eDNA is necessary for the initial establishment of *P. aeruginosa* biofilms [12, 13]. Filamentous networks of eDNA were shown to stabilize the biofilm architecture. EPS lipids contribute to the hydrophobic properties of EPS [14].

The biofilm is enriched with specific protein adhesins that mediate known molecular binding mechanisms for irreversible attachment. In addition, membrane transport proteins such as porins and extracellular enzymes are up-regulated [15]. Biofilm formation occurs in a sequential process of: (i) transport of microbes to a surface by chemotaxis or Brownian motion; (ii) initial attachment; (iii) irreversible attachment of bacteria and formation of microcolonies; (iv) biofilm maturation; and (v) detachment [16].

The substratum characteristics may influence biofilm formation and morphology. Most investigators have found that microorganisms attach more rapidly to hydrophobic, nonpolar surfaces such as teflon and other plastics than to hydrophilic materials such as glass or metals [17]. Furthermore, microbial colonization increases with the increase in surface roughness [18]. The biofilm architecture changes constantly, due to external and internal processes [16]. The biofilm thickness may be affected by the number and species of microorganisms. Biofilms of pure cultures of either *Klebsiella. pneumoniae* or *P. aeruginosa* in a laboratory reactor were thinner (15 μ and 30 μ, respectively), whereas a biofilm containing both species was thicker (40 μ) [19]. Mature bacterial biofilms can adopt various architectures depending on the characteristics of the surrounding environment, such as nutrients, pH, temperature, shear forces, osmolarity and composition of the microbial consortia. The common complex biofilm is a mushroom-like structure which is surrounded by highly permeable water channels that facilitate the transport of nutrients and oxygen to the interior of the biofilm [8].

http://dx.doi.org/10.5772/58432

Cell surface-associated proteins such as pili, flagella, curli and amyloid fibers are believed to be important factors for biofilm formation [20]. Cell-to-cell signaling (quorum sensing) has been demonstrated to play a role in biofilm formation. *P. aeruginosa* produces two different quorum sensing molecules, lasR-lasI and *rhlR-rhlI,* which were shown to be involved in biofilm formation [21]. The biofilm of a double mutant produced thinner biofilms than the wild type, its cells were more densely packed, and the typical biofilm architecture was absent. Addition of the quorum sensing molecules known as homoserine lactone to a medium that contained the mutant biofilms resulted in biofilms with a structure and thickness similar to that of the wild type [17,22].

In conclusion, knowledge and understanding of the biofilm components, morphology and formation steps may lead to better biofilm conductivity in microbial fuel cells (MFC).

4. Spectroscopic methods for anode biofilm study, including advancing real-time analysis

During the last decade, most efforts in microbial fuel cell (MFC) research focused on modifications of MFC design and electrode materials, with little investigation of the properties of the anode microorganisms that are essential for maximal current production. However, understanding the functions of microbial cell surfaces requires knowledge of their chemical structural, physical properties and biological processes.

Perhaps the most challenging effort to improve MFC power production lay in the fundamental understanding of the biofilm's chemical, physical and biological characteristics. Study of the local properties of an anode biofilm is an even a greater challenge, due to the low concentration of bacterial cells. In addition to conventional methods such as scanning electron microscopy (SEM) and confocal laser scanning microscopy (CLSM), there exist modern methods such as Raman spectroscopy, Fourier transform infrared spectroscopy (FTIR) and atomic force microscopy (AFM). These latter are nondestructive methods which can probe biofilms down to single-cell surfaces with high resolution, and thereby promote better understanding of biofilm formation, functionality and activity.

In this chapter we will discuss the principle, the advantages and the disadvantages of each of the methods and the application of these techniques in MFCs.

4.1. Confocal laser scanning microscopy

Confocal laser scanning microscopy (CLSM) is often used in optical imaging of sliced microfluidic velocity fields by mapping the investigated focal plane. CLSM enables obtaining a series of optical sections of intact undisturbed biological samples as thin as 0.3 μm. Commonly used analyses that rely on staining techniques are applied to determine the architecture, spatial distribution and viability profile in microbial biofilms. The most popular application of CLSM is for identification of live and dead bacteria. Simultaneous measurements of anodic biofilms during the MFC's operation may be obscured by the need to apply labeling materials. This method is therefore used more commonly as a useful *ex situ* method [23, 24].

4.2. Scanning electron microscopy

Scanning electron microscopy (SEM) analysis provides an excellent magnification technique that uses a condensed electron beam to scan a sampled object and magnify specific regions of its surface area. Highly resolved images can be produced by SEM to provide morphological details of the surface, information on the three-dimensional topography as well as an elemental composition analysis map of the same unit area when coupled with an x-ray elemental detection sensor. SEM is therefore instrumental in a broad spectrum of scientific and industrial applications.

The main shortcomings of SEM are the relatively low resolution compared to transmission electron microscopy (TEM), which is usually higher than a few tens of nanometers and the need for a surface-conducting coating layer to avoid local charging and heating effects. Moreover, SEM is an ultra-low pressure technique. All samples must therefore withstand the low pressure inside the SEM vacuum chamber [25].

4.3. Raman spectroscopy

The Raman spectroscopy method is based on inelastic scattering of monochromatic photons, usually from a laser source. In this process, the frequency of the back-scattered photons changes their frequency due to interaction with a sample. The process begins when photons of the laser light are absorbed by the sample and are then re-emitted at a longer wavelength compared with the laser's original monochromatic frequency. This phenomenon is called the Raman frequency shift effect. This shift provides information on vibrational, rotational and other low frequency transitions at the molecular level. Raman spectroscopy is typically applied in the study of solid, liquid and gaseous samples.

The main strength of the Raman technique is its high sensitivity, the fact that it does not require any staining and its ability to generate detailed chemical and spatial information with a resolution below the diffraction limit. A combination of Raman spectroscopy with optical microscopy provides a powerful source of information which is both detailed and sensitive and can be obtained with a spatial resolution below the diffraction limit [26, 27]. The main disadvantages of the Raman technique is the weak Raman effect, which requires sensitive and highly optimized instrumentation for detection. Furthermore, fluorescence of the sample or impurities within it can mask the Raman spectrum. Sample heating through the intense laser radiation can damage the sample and distort the Raman spectrum [25].

4.4. Atomic force microscopy

Atomic force microscopy (AFM) provides one of the highest topographical profile imaging of a sampled surface, sometimes even on a nanometric scale. It does so by measuring interaction or repulsion forces between a sharp nano-size probe and a surface. The nitration distances between the tip and the sample can be as small as a few tenths of a nanometer. The AFM tip can be used either in a contact mode where the tip actually touches the surface, or in a non-contact mode where van der Waals interactions produce forces across the short distance between the tip and the surface. The forces produced by each mode of operation can be recorded to give topographic, conductive, magnetic, mechanical (friction forces) or potential images of the surface.

In fact, AFM imaging does not require any surface modification that may damage or change the sample. Most importantly, AFM techniques are well suited to work in ambient air or in a liquid environment, including physiological conditions [25]. Therefore, the study of electrodes, biological samples, biomolecules and even living organisms can be measured rather easily by AFM. Moreover, advanced surface-imaging tools offer more than just force measurements. It can be applied to probe physical properties such as molecular interactions, surface hydrophobicity, surface charges, mechanical properties and local electrochemical properties [28]. For example, AFM enables imagining of biofilms and the EPS distribution within the biofilm [29].

The disadvantage of AFM is that the scanned area of the AFM image is rather small (few tens of microns), and it is very sensitive to externally generated electrical and mechanical noises [25].

5. Power-producing bacterial cells

Power-producing bacterial species that have exoelectrogenic activity without exogenous mediators include: *Shewanella putrefaciens, Clostridium butyricum, Desulfuromonas acetoxidans, Geobacter metallireducens, Geobacter sulfurreducens, Rhodoferax ferrireducens, Pseudomonas aeruginosa, Desulfobulbus propionicus, Geothrix fermentans, Shewanella oneidensis, Escherichia coli, Rhodopseudomonas palustris, Ochrobactrum anthropi, Desulfovibrio desulfuricans, Acidiphilium sp. Klebsiella pneumonia, Thermincola sp.* [30] and *Cupriavidus basilensis* [31]. Microorganisms which were detected on the anode of microbial electrolysis cells (MECs) include: *Stenotrophomonas, Lactobacillus, Curtobacterium, Agrobacterium, Flavobacterium, Bradyrhizobium, Pseudomonas, Desulfovibrio, Shewanella, Desulfonauticus, Xenohaliotis* and *Marinicola* [32].

Pure cultures are suitable for basic study of MFCs, since they allow analysis of electrochemical and biochemical processes as well as high reproducibility. However, mixed cultures are more suitable for industrial applications because no sterilization is required and it is not necessary to maintain anaerobic conditions via an external N_2 flow. Mixed bacteria usually produce higher power densities in MFCs than pure bacterial strains. It is important to note that the power density depends on the specific MFC configuration, electrode spacing, bacterial cell species, growth medium and the cathode catalysts [1, 2]. Thus, power densities produced in a MFC using a specific bacterium cannot be directly compared with another bacterium unless all other parameters are identical.

Exoelectrogenic bacteria transfer electrons to anodes either directly or via self-produced mediators. In the direct mechanism, electron transfer occurs via membrane-associated C-cytochrome or through conductive pili or appendages [33]. In mediated mechanisms, electron transfer between the bacterial cell and the anode surface occurs through self-produced soluble redox compounds such as flavins or pyocyanin [34]. Well-known exoelectrogens include *Geobacter* sp. and *Shewanella* sp. [35] which are known to have outer-membrane cytochromes and conductive pili, and *Pseudomonas* sp. that produce mediators that shuttle electrons from the bacteria through the biofilm matrix to the anode [36].

6. Mechanisms of electron transfer from the biofilm to the anode

6.1. C-cytochrome

C-type cytochromes have been considered as one of the most important electron transfer strategies in current generation by exoelectrogens. C-cytochromes are heme-containing proteins which are widespread in most bacteria. *S. oneidensis* MR-1 has 42 putative C-cytochromes and 80% of them are located in the outer membrane, covering 8-34% of the cell surface. It was reported that in *Geobacter*, C-cytochromes are bound to the matrices of extracellular polymeric substrates [37]. Appendages termed bacterial nanowires that were identified in both *Shewanella* and *Geobacter* also contain surface-located C-cytochromes which are postulated to transport electrons from distant cells to electrodes [38, 39]. However, these studies are based on *in vitro* experiments and more research needs to be done to understand this mechanism [40]. *Geobacter* biofilm respiration was found to continue after the interruption of electrode polarization, since these bacteria can store electrons in the heme of the exocytoplasmic cytochromes. When the electrode was connected again, stored electrons were recovered as a current superimposed on the basal steady-state current [41].

Cytochrome genes were examined during extracellular electron transport in MEC. A total of 21 cytochrome genes were detected. Four bacterial genera contain the cytochrome genes: *Geobacter, Desulfovibrio, Rhodopseudomonas* and *Shewanella,* all of which increased over three months of the MEC reactor's performance [32]. Microarray analysis of *Geobacter sulfurreducens* thick anode biofilms (>50 μm) in MFCs revealed 13 genes in current-harvesting biofilms. Up-regulated genes included two outer C-type membrane cytochromes, OmcB and OmcZ. Down-regulated genes included the genes for the outer-membrane C-cytochromes OmcS and OmcT. Results of quantitative RT-PCR of gene transcript levels during biofilm growth were consistent with microarray results. OmcZ and the outer-surface C-type cytochrome OmcE were more abundant and OmcS was less abundant in current-harvesting cells. The role of outer-surface proteins whose genes were expressed in the current-producing biofilm versus the control biofilm which reduced fumarate were evaluated by gene deletion and its impact on current production was examined. It was found that deletion of OmcS, OmcB, or OmcE had no impact on maximum current production. Deletion of OmcZ significantly reduced power production. These results suggest that OmcZ is a key component in electron transfer in MFCs with a *G. sulfurreducens* anode [42]. In thin (ca. 10 mm) wild type biofilms, genes for OmcS and OmcE are more highly expressed than in planktonic cells grown with a soluble electron acceptor such as Fe(III) citrate [43]. CymA, which participates in many *Shewanella* anaerobic respiration processes, is a tetraheme C-cytochrome whose C-terminal is exposed to the periplasm and its N-terminal is anchored in the inner membrane. In the case of electrode reduction, a deletion of the CymA gene caused >80% decrease in current generation. CymA can interact directly with many terminal reductases in the periplasm, such as fumarate reductase and nitrate reductase. Bacterial two-hybrid showed pair-wise interactions among CymA, MtrA and some other periplasmic redox proteins, indicating that CymA is the major electron conduit to the periplasmic space and can interact directly with periplasmic redox proteins by forming a transient protein complex [44].

Okamoto et al identified voltammetry signals of outer membrane C-cytochrome in monolayer biofilms of *S. oneidensis* [45]. Okamoto et al analyzed monolayer and multilayer biofilms of *S. oneidensis* on tin-doped indium oxide electrodes and compared the respective amounts of electroactive C-cytochrome using voltammetry techniques. The scan-rate dependence of cyclic voltammograms was used to investigate the role of C-cytochrome in the transfer of respiratory electrons to distant tin-doped indium oxide electrodes. Electron conduction in *S. oneidensis* MR-1 multilayer biofilms was demonstrated to be mediated by sequential redox cycling of outer membrane C-cytochrome under normal physiological conditions. It was also demonstrated that the electron transport outer membrane C-cytochrome conduit across the biofilms contributed to the anodic current generation [40].

Immunogold labeling of the outer-surface C-type cytochrome OmcZ showed that when *G. sulfurreducens* grew as a biofilm on a graphite electrode that served as an anode and sole electron acceptor for growth, OmcZ was highly concentrated at the biofilm–electrode interface. Control biofilms which were grown on the same graphite material but with fumarate as the electron acceptor did not have OmcZ accumulations at the biofilm–electrode interface. The researchers suggested that OmcZ may serve as an electrochemical gate facilitating electron transfer from *G. sulfurreducens* biofilms to the anode surface [46]. Direct current surface-enhanced infrared (IR) absorption spectroscopy and FTIR demonstrated a linear correlation between the increasing presence of *Geobacter sulfurreduces* protein and current production. This result may confirm that the extracellular cytochromes [47] are responsible for the electron transfer to the gold electrode [48].

Eaktasang et al. electrochemically oxidized the surface of a graphite felt electrode with a strong acid in order to stimulate current production in the MFC. FT-IR was used to examine the chemical property changes in the graphite felt surfaces which were induced by its electrochemical oxidation using acid treatment. Current production in the MFC equipped with the surface-modified graphite anode was about 40% higher than that obtained from the MFC of a bare graphite anode. FT-IR spectra of surface-modified felt and bare graphite felt showed a notable broad band at 3264 cm^{-1} ascribed to stretching vibration of OH within –COOH, a broad peak at 1560cm^{-1} ascribed to C=O of ketone and carboxyl groups, and a peak at 1419 cm^{-1} ascribed to bending vibration of –OH. The relative intensities of the abovementioned peaks increased significantly after electrochemical oxidation treatment of graphite felt, indicating that alcohol and carboxyl functional groups formed on the surface of the graphite-felt hydrogen bonding with peptide bonds in the bacterial cytochrome. They also demonstrated that the carboxylic acid terminus of gold-modified electrodes can facilitate the binding with cytochrome on its surface, and as a result, current production increased due to the enhanced transfer of electrons from the interior of the cell [49]. *In vitro* and *in vivo* experiments led to the proposal of several models based on the redox properties of bacterial C-cytochromes. A 4-step mechanism for electron transfer in *Geobacter* biofilms was proposed: step 1, acetate uptake and electron transfer to periplasmic cytochromes; step 2, subsequent electron transfer to the exocytoplasmic cytochromes; step 3, electron transport along the biofilm matrix cytochromes; and step 4, transfer between the interfacial cytochromes and the electrode [41].

6.2. Pili

Genes encoding proteins with a PilZ domain were deleted from the G. sulfurreducens genome in an attempt to study the importance of pili to biofilm conductivity. The mutant strain designated as strain CL-1 produced more pili than the wild type strain and formed 6-fold more conductive biofilms than the wild type strain. Heme-staining revealed a higher abundance of cytochrome with a molecular weight consistent with OmcS in CL-1 and Western blot analysis with OmcS-specific antiserum confirmed higher production of OmcS in CL-1. Immunogold labeling coupled with TEM demonstrated that OmcS was localized on the pili of CL-1 [50]. Multilayer biofilms of *Geobacter sulfurreducens* on the anode surface of a MFC remained viable even at a distance from the anode. There was no decrease in the efficiency of current production with an increase in the thickness of the biofilm. Genetic studies demonstrated that efficient electron transfer through the biofilm required the presence of electrically conductive pili which represent an electronic network promoting long-range electrical transfer in an energy-efficient manner in the MFC [38].

G. sulfurreducens KN400 bacterial cells which were selected after 5 months of incubation at-400 mV produced higher current and power densities than the original inoculum (strain DL1). The enhanced capacity for current production in KN400 was associated with a greater abundance of electrically conductive microbial nanowires than DL1 and lower internal resistance in KN400 fuel cells. KN400 produced flagella, whereas DL1 does not. The changes in outer surface components were associated with a greater propensity of strain KN400 to stick to surfaces than strain DL1. KN400 cells grown with fumarate as the electron acceptor were clumpy and strongly adhered to the glass surface of the culture tubes. This research showed that microorganisms' ability to electrochemically interact with electrodes can be enhanced with the appropriate selective pressure and that improved current production is associated with clear differences in the properties of the outer surface of the cell that may provide better microbe–electrode interactions [51].

The AFM technique, together with electrochemical measurements by scanning tunneling microscopy (STM), allowed mapping of conducting substrates and microbial pili in MFCs studies. These two techniques enabled identification of the pili on *Geobacter* species. The existence of such pili raises the possibility of long-range direct electron transfer from each microbe to the electrode [36, 48]. El-Naggar et al. evaluated transport along the bacterial nanowires by conducting probe AFM at several points alongside a single nanowire using a metallic electrode and a conductive AFM tip. *S. oneidensis* MR-1 nanowires were electrically conductive along micrometer-length scales, yielding a corresponding electron transport rate, at 100 mV, of about 109 electrons per second, with an estimated resistance of 1 Ω (Fig 1). It was also found that mutants deficient in genes for the C-type decaheme-cytochromes MtrC and OmcA produced supplements that are morphologically consistent with bacterial nanowires, but were found to be nonconductive [52].

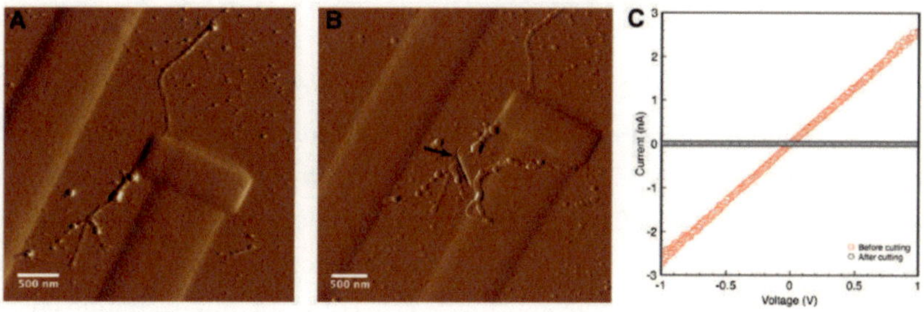

Figure 1. Measuring electrical transport along a bacterial nanowire. (A) Tapping-mode atomic force microscopy (AFM) amplitude image detailing the contact area with the bacterial nanowire from Fig. 1. (B) Contact-mode AFM deflection image of the junction after cutting the nanowire with FIB milling. The arrow marks the cut location. (C) Current-voltage curve of the bacterial nanowire (ramp-up and ramp-down) both before (red) and after (black) cutting the nanowire [52].

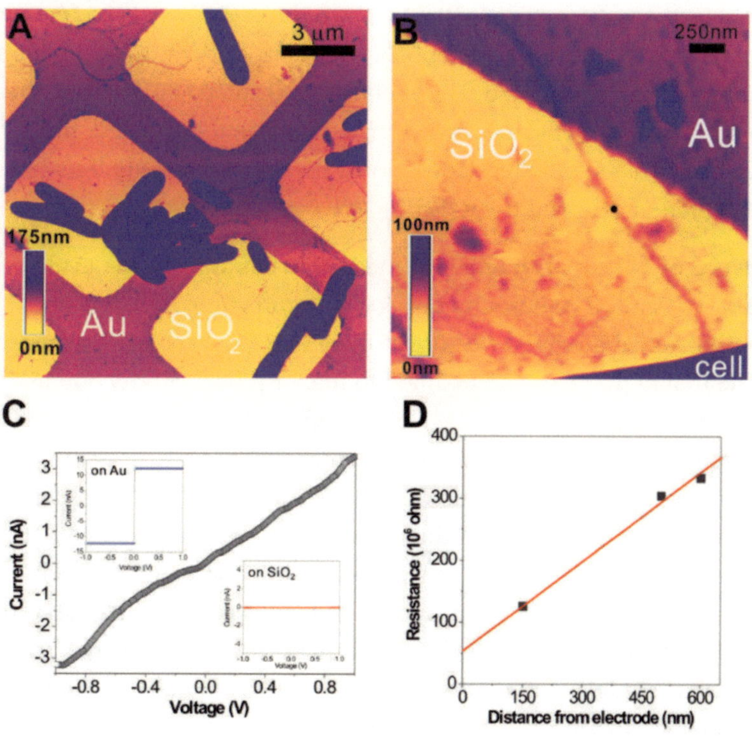

Figure 2. CP-AFM of a bacterial nanowire. (A) Topographic AFM image showing air-dried S. oneidensis MR-1 cells and extracellular appendages deposited randomly on a SiO2/Si substrate patterned with Au microgrids. (B) Contact mode AFM image showing a nanowire reaching out from a bacterial cell to the Au electrode. (C) An I-V curve obtained by probing the nanowire at a length of 600 nm away from the Au electrode (at the position marked by the black dot in B). (Inset) The I-V curves obtained on bare Au and SiO2, respectively. (D) A plot of total resistance as a function of distance between AFM tip and Au electrode [52].

7. The effect of medium and pH conditions on power production

The effects of anodic pH on electricity production in two-chamber MFCs inoculated with anaerobic activated sludge were examined. The maximum power density was 1170±58 mWm^2 with a current density of 0.18 $mA \times cm^{-2}$ at pH 9.0, which was 29% and 89% higher than in MFCs operated at pH 7.0 and 5.0, respectively. Electrochemical measurements demonstrated that the environmental pH may influence the electron transfer kinetics of the anodic biofilms. At pH 9.0, the apparent electron transfer rate constant and exchange current density were greater, whereas charge transfer resistance was smaller than under other pH conditions. SEM analysis reveled better biofilm formation at pH 9.0 compared to pH 7.0 and 5.0. The biofilm at pH 9.0 showed increased electron transfer efficiency with respect to the electrocatalytic current, electron transfer rate, exchange current density, and charge transfer resistances, compared with biofilms at pH 7.0 and 5.0. These results demonstrate that electrochemical interactions between bacteria and electrodes in MFCs are greatly enhanced under alkaline conditions, which can an important reason for the improved current production in MFCs [53]. It should be noted that the cathodic potentials were almost identical under all pH conditions, whereas the anodic potentials varied. The anode that operated at pH 9.0 had the most negative individual potentials in the entire current range, whereas the anode at pH 5.0 had the most positive individual potentials. The variations in the power outputs thus resulted from the anodes under different pH conditions [53].

The performance of two dual-chambered mediators-less MFCs was evaluated at different sludge loading rates and pH environments. A maximum volumetric power of 15.51 W/m^3 and 36.72 W/m^3 was obtained in MFC-1 (feed pH 6.0) and MFC-2 (feed pH 8.0), respectively. These results indicate that higher feed pH (8.0) led to higher power production [54]. Zhuang et al. discovered that an alkaline anode (pH 10) led to higher power production during wastewater treatment compared to a MFC which worked at neutral pH. Their assumption is that there methanogenesis is suppressed at higher pHs, and this contributes to a significant enhancement of coulombic efficiency [55]. In contradistinction to the abovementioned studies, there are several studies which described that higher power production was obtained in natural or acidic environments. An air-cathode MFC which operated with a mixed bacterial culture at different pHs showed that the anodic microbial process preferred a neutral pH and that microbial activities decreased at higher or lower pHs [56]. A MFC with a pure culture of *Enterobacter cloacae* was evaluated as a function of variations in the pH microenvironment. Operation under pH 6.5 and 7.4 led to maximum current generation of 0.40 mA and 0.42 mA, respectively. However, MFC operation under higher pH environments of 8.5 and 9.5 led to a maximum current generation of only 0.38 mA and 0.27 mA, respectively [57].

In conclusion, we assume that MFC power production is dependent on the pH environment and may correlate with the bacterial cell anode species.

8. Anode biofilm thickness, morphology, viability and conductivity

CSLM analyses revealed a correlation between current and increasing coverage of the anode surface with cells. The average height of the biofilm pillars at currents nearing the maximum current outputs in batch mode was 40 μm (±6 μm), and some cell clusters were as high as 50 μm. Protein measurements of the biofilm biomass at different points during current production indicated a direct linear increase in the amount of biomass on the anodes as the current increased and the biofilms developed. Viability staining indicated that during biofilm development, cells at a distance from the electrode surface remained viable, metabolically active and involved in electron transfer to the anode [38].

Geobacter can form multi-microbe thick, more than 20-cell length, persistent biofilms when the anode serves as their terminal electron acceptor. This phenomenon was not observed in the presence of insoluble oxidants for respiration. The combination of robust direct electron transfer and high cell surface density enables *Geobacter* biofilms to achieve higher anodic current densities than any other species [58].

Three MFCs (plain graphite electrodes, air cathode, Nafion membrane) were operated separately with variable biofilm coverage [control; anode surface coverage (0%), partially developed biofilm (coverage ~44%) and fully developed biofilm (coverage of ~96%)] under acidophilic conditions (pH 6) at room temperature. Higher specific power production [29 mW/kg CODR (CW and DSW)], specific energy yield [100.46 J/kg VSS (CW)], specific power yield [0.245 W/kg VSS (DSW); 0.282 W/kg VSS (CW)] and substrate removal efficiency of 66.07% (substrate degradation rate, 0.903 kg COD/m^3-day) were observed, especially with fully developed biofilm operation [59].

The behavior of MFCs during initial biofilm growth and characterization of anodic biofilm were studied using two-chamber MFCs with activated sludge as inoculum. When the biofilms were well developed, a maximum closed circuit potential of 0.41 V and 0.37 V (1000 Ω resistor) was achieved using acetate and glucose, respectively. SEM analysis revealed rod-shaped cells, 0.2-0.3 mm wide by 1.5-2.5 mm long, in the anode biofilm in the acetate-fed MFC, mainly arranged in a monolayer. The biofilm in the glucose-fed MFC was made of cocci-shaped cells in chains and a thick matrix [60. A MFC was operated with a pure culture of *Cupriavidus basilensis* bacterial cells grown in a defined medium containing acetate or phenol. Operating this mediator-less MFC under a constant external resistor of 1 kΩ with acetate or phenol led to a current generation of 902 and 310 mA m^{-2}, respectively. SEM and confocal microscopy analyses demonstrated a developed biofilm with pili/appendages and a live anode *C. basilensis* population which was stained with a LIVE/DEAD viability kit and further analyzed by CSLM (Fig 3) [31].

Electron transfer between bacterial cells and the electrode is one of the major blockages to a desired power density. A direct electron transfer process between *P. aeruginosa* bacterial cells and the electrode was investigated using cationic reagents which are known to "perforate" the bacterial membrane. Three reagents, chitosan, ethylenediaminetetraacetic acid (EDTA) and polyethyleneimine (PEI), were explored. The surface morphologies of chemically treated

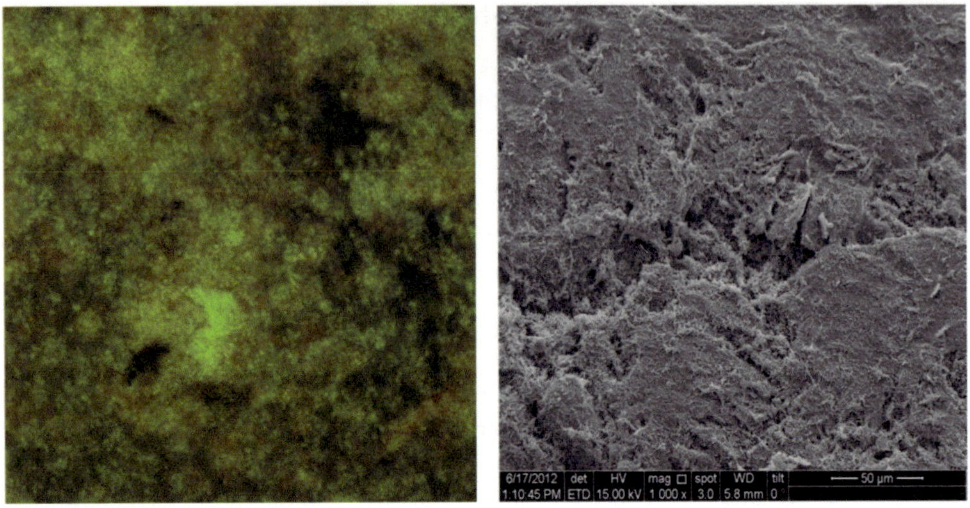

Figure 3. *Cupriavidus basilensis* biofilm grown on a graphite anode in MFC. CSLM micrograph, magnification of 40 (A). SEM micrograph, magnification of 1,000 (B).

P. aeruginosa cells were examined with SEM and AFM images. The results showed that the chemical treatment caused changes in the cell shape and perforated bacterial cells, contrary to the original *P. aeruginosa* cells that were plump and smooth. The cells were flattened with pores and clusters on the surface. The chemical treatment disorganized the outer membrane of *P. aeruginosa* which resulted in higher permeability, like "drilling", which significantly improved electron transfer for high power density MFCs. Cyclic voltammograms (CV) of the anaerobically activated control and perforated *P. aeruginosa* cells were measured in glucose-free medium and showed two pairs of redox peaks at-0.59 and-0.47 V. The two redox waves represented different electron transfer mechanisms of this bio-anode. The output current profiles at a constant load of chitosan, EDTA-and PEI-treated cells were much higher than in control cells, and the PEI-treated cell catalysts produced the highest discharge (oxidation) current density, which parallels the result of the CV experiment. The polarization curves revealed that the PEI-treated cell has much lower polarization for better energy conversion efficiency than the control cells. It was explained that the increased permeability of the bacterial outer membrane by the large pores and channels facilitates the transport of redox mediators and catabolic enzymes through the cell membrane and thus provides faster electron transfer [61].

Introducing a pH-sensitive fluoroprobe into the anode chamber revealed a strong pH gradient within the *G. sulfurreducens* anode biofilm. The pH decreased with increased proximity to the anode surface and from the exterior to the interior of the biofilm pillars. pH levels near the anode surface were as low as 6.1 compared to pH 7 near the cathode. Various controls demonstrated that proton accumulation was associated with current production. Decreasing the pH of the culture medium from 7 to 6 limited the growth of *G. sulfurreducens*. The results demonstrated that it is feasible to non-destructively monitor the activity of anode biofilms in real time. It was also suggested that the accumulation of protons that are released from organic matter oxidation within anode biofilms can limit current production [62].

Millo et al. employed surface-enhanced resonance Raman (SERR) spectroscopy in combination with CV analysis. This technique, performed under strict electrochemical control, reveals the redox, coordination and spin states of the heme iron in addition to the nature of its axial ligand. Furthermore, by applying *in situ* SERR spectroscopy electrochemical analysis on a catalytically active biofilm grown on silver electrodes, it was revealed that two bis (histidine) coordinated heme cytochrome redox couples are involved in the mediated electron transfer between the bacteria and the electrode [63].

Liu et al. showed that graphene modification improved power density and energy conversion efficiency by almost 3 times that of a *P. aeruginosa* mediator-less MFC. Raman analysis revealed that the improvement was credited to the high biocompatibility of graphene which promotes bacterial growth on the electrode surface and results in the creation of more direct electron transfer activation centers and stimulates excretion of mediating molecules for higher electron transfer [65]. Raman spectroscopy analysis at different growth stages revealed changes in the vibrational properties of *Geobacter* Cyt *c* resulting from shifts in the anodic potential between different redox conditions [65].

In conclusion, the biofilm strain composition, components, morphology and thickness play a major role in the electrochemical process in MFCs, and show marked influence on bioelectricity production. Changes in biofilm parameters may influence the anode electrochemical characteristics and performance in a MFC. It is therefore important to understand the characteristics, structure and composition of the biofilm in order to ensure optimized operation of MFCs.

Acknowledgements

This chapter was supported in part by the Research Authority of Ariel University, the Israel Ministry of Environmental Protection and the Samaria and Jordan Rift Valley Regional R&D Center.

Author details

Michal Schechter[1], Alex Schechter[2], Shmuel Rozenfeld[1], Emanuel Efrat[1] and Rivka Cahan[1*]

*Address all correspondence to: rivkac@ariel.ac.il

1 Department of Chemical Engineering, Ariel University, Ariel, Israel

2 Department of Biological Chemistry, Ariel University, Ariel, Israel

References

[1] Rabaey K., Boon N., Hofte M., Verstraete W. Microbial Phenazine Production Enhances Electron Transfer in Biofuel Cells. Environmental Science and Technology 2005; 39(9) 3401–3408.

[2] Liu BF., Ren NQ., Tang J., Ding J., Liu WZ., Xu JF., Cao GL., Guo WQ., Xie G.J. Bio-Hydrogen Production by Mixed Culture of Photo-and Dark-Fermentation Bacteria. International Journal of Hydrogen Energy 2010; 35(19) 2858–2862.

[3] Donlan RM., Biofilms: Microbial Life on Surfaces. Emerging Infectious Diseases Journal 2002; 8(9) 881-890.

[4] Branda SS., Vik S., Friedman L., Kolter R. Biofilms: the Matrix Revisited. Trends Microbiology. 2005; 13(1) 20-26.

[5] Flemming HC., Neu TR., Wozniak DJ. The EPS Matrix: The House of Biofilm Cells. Journal of Bacteriology. 2007; 189(22) 7945-7947.

[6] Sutherland IW. The Biofilm Matrix-an Immobilized but Dynamic Microbial Environment. Trends Microbiology. 2001; 9(5) 222-227.

[7] Flemming HC., Wingender J., Extracellular Polymeric Substances (EPS): Structural, Ecological and Technical aspects, in Encyclopedia of environmental microbiology, Bitton G, Editor. John Wiley & Sons: New York. 2002; p. 1223-1231.

[8] Kolter R., Greenberg EP. Microbial Sciences: the Superficial Life of Microbes. Nature. 2006; 441(7091) 300-302.

[9] Leriche V., Sibille P., Carpentier B. Use of an Enzyme-Linked Lectinsorbent Assay to Monitor the Shift in Polysaccharide Composition in Bacterial Biofilms. Applied and Environmental Microbiology 2000; 66(5) 1851–1856.

[10] [10] Celmer D., Oleszkiewicz JA, Cicek N. Impact of Shear Force on the Biofilm Structure and Performance of a Membrane Biofilm Reactor for tertiary Hydrogen-Driven Denitrification of Municipal Wastewater. Water Research. 2008; 42(12) 3057-3065.

[11] Gieseke A., Arnz P., Amann R., Schramm A. Simultaneous P and N Removal in a Sequencing Batch Biofilm Reactor: Insights from Reactor-and Microscale Investigations. Water Research. 2002; 36(2) 501-509.

[12] Allesen-Holm M., Barken KB., Yang L., Klausen M., Webb JS., Kjelleberg S., Molin S., Givskov M., Tolker-Nielsen T. A Characterization of DNA Release in *Pseudomonas aeruginosa* Cultures and Biofilms. Molecular Microbiology. 2006; 59(4) 1114–1128.

[13] Whitchurch CB., Tolker-Nielsen T., Ragas PC., Mattick JS. Extracellular DNA Required for Bacterial Biofilm Formation. Science. 2002; 295 1487.

[14] Conrad A., KontroM., Keinänen MM., Cadoret A., Faure P., Mansuy-Huault L., Block JC. Fatty Acids of Lipid Fractions in Extracellular Polymeric Substances of Activated Sludge Flocs. Lipids. 2003; 38(10)1093-1105.

[15] Dunne WM. Bacterial Adhesion: Seen Any Good Biofilms Lately? Clinical Microbiology Reviews. 2002; 15(2) 155-166.

[16] Tolker-Nielsen T., Brinch UC., Ragas PC., Andersen JB., Jacobsen CS., Molin S. Development and Dynamics of *Pseudomonas* sp. Biofilms. Journal of Bacteriology. 2000; 182(22) 6482–6489.

[17] Fletcher M., Loeb GI. Influence of Substratum Characteristics on the Attachment of a Marine *Pseudomonad* to Solid Surfaces. Applied and Environmental Microbiology. 1979; 37(1) 67–72.

[18] Characklis WG., McFeters GA., Marshall KC. Physiological Ecology in Biofilm Systems. In: Characklis WG, Marshall KC, editors. Biofilms. New York: John Wiley & Sons; 1990. p. 341–94.

[19] James GA., Beaudette L., Costerton JW. Interspecies Bacterial Interactions in Biofilms. Journal of Industrial Microbiology. 1995;15(4) 257-62.

[20] Gohl O., Friedrich A., Hoppert M., Averhoff B. The Thin Pili of *Acinetobacter* sp. Strain BD413 Mediate Adhesion to Biotic and Abiotic Surfaces. Applied and Environmental Microbiology. 2006; 72(2) 1394-1401.

[21] Davies DG., Parsek MR., Pearson JP., Iglewski BH., Costerton JW., Greenberg EP. The Involvement of Cell-to-Cell Signals in the Development of a Bacterial Biofilm. Science. 1998; 280(5361) 295–298.

[22] Xie H., Cook GS., Costerton JW., Bruce G., Rose TM., Lamont RJ. Intergeneric Communication in Dental Plaque Biofilms. Journal of Bacteriology. 2000; 182 7067–7069.

[23] Astner S., Ulrich M. Confocal Laser Scanning Microscopy. Hautarzt. 2010; 61 421-428.

[24] Murphy DB., Davidson MW. Confocal Laser Scanning Microscopy. In: Fundamentals of Light Microscopy and Electronic Imaging. 2012; 265-305.

[25] Niemantsverdriet JW. Spectroscopy in catalysis: WILEY-VCH Verlag GmbH & Co. kgAa.2007.

[26] Neugebauer U., Rösch P., Schmitt M., Popp J., Julien C., Rasmussen A., Budich C., Deckert V. On the Way to Nanometer-Sized Information of the Bacterial Surface by Tip-Enhanced Raman Spectroscopy. ChemPhysChem. 2006; 7(7) 1428-1430.

[27] Pradhan N., Pradhan SK., Nayak BB., Mukherjee PS., Sukla LB., Mishra BK. Micro-Raman analysis and AFM imaging of *Acidithiobacillus ferrooxidans* biofilm grown on uranium. Research in Microbiology. 2008; 159(7-8) 557-561.

[28] Dufrêne YF. Atomic Force Microscopy of Fungal Cell Walls: an Update. Yeast 2010; 27(8) 465-471.

[29] Beech IB., Smith JR., Steele AA., Penegar I., Campbell SA. The Use of Atomic Force Microscopy for Studying Interactions of Bacterial Biofilms with Surfaces. Colloids and Surfaces B: Biointerfaces. 2002; 23 231-247.

[30] Logan BE. Exoelectrogenic Bacteria that Power Mmicrobial Fuel Cells. Nature Reviews. 2009; 7 375-381.

[31] Friman H., Schechter A., Ioffe Y., Nitzan Y., Cahan R. Current Production in a Microbial Fuel Cell Using a Pure Culture of *Cupriavidus basilensis* Growing in Acetate or Phenol as a Carbon Source. Microbial Biotechnology. 2013; 6(4) 425-34.

[32] Liu W., Wang A., Sun D., Ren N., Zhang Y., Zhou J. Characterization of Microbial Communities During Anode Biofilm Reformation in a Two-Chambered Microbial Electrolysis Cell (MEC). Journal of Biotechnology. 2012; 157(4) 628–632.

[33] Mehta T., Coppi MV., Childers SE., Lovley DR. Outer Membrane C-Type Cytochromes Required for Fe(III) and Mn(IV) Oxide Reduction in *Geobacter sulfurreducens*. Applied and Environmental Microbiology 2005; 71(12) 8634-8641.

[34] Enrico M., Baron DB., Shikhare ID., Coursolle D., Gralnick JA., Bond DR. *Shewanella* Secretes Flavins that Mediate Extracellular Electron Transfer. Proceedings of the National Academy of Sciences. 2008;105(10) 3968-3973.

[35] Kim JR., Jung SH., Regan JM., Logan BE. Electricity Generation and Microbial Community Analysis of Alcohol Powered Microbial Fuel Cells. Bioresource Technology. 2007; 98 2568–2577.

[36] Reguera G., McCarthy KD., Mehta T., Nicoll JS., Tuominen MT., Lovley DR.. Extracellular Electron Transfer via Microbial Nanowires. Nature 2005; 435(7045) 1098–1101.

[37] Richter H., Nevin KP., Jia H., Lowy DA., Lovley DR. Tender LM. Cyclic Voltammetry of Biofilms of Wild Type and Mutant *Geobacter sulfurreducens* on Fuel Cell Anodes Indicates Possible Roles of OmcB, OmcZ, Type IV Pili, and Protons in Extracellular Electron Transfer. Energy and Environmental Science. 2009; 2 506–516.

[38] Reguera G., Nevin KP., Nicoll JS., Covalla SF., Woodard TL., Lovley DR. Biofilm and Nanowire Production Leads to Increased Current in *Geobacter sulfurreducens* Fuel Cells. Applied and Environmental Microbiology. 2006; 72(11) 7345–7348.

[39] Gorby YA., Yanina S., McLean JS., Rosso KM., Moyles D., Dohnalkova A., Beveridge TJ., Chang IS., Kim BH., Kim KS., Culley DE., Reed SB., Romine MF.,Saffarini DA., Hill EA., Shi L., Elias DA., Kennedy DW., Pinchuk G., Watanabe K., Ishii S., Logan B., Nealson KH., Fredrickson JK. Electrically Conductive Bacterial Nanowires Produced by *Shewanella oneidensis* Strain MR-1 and Other Microorganisms, Proceedings of the National Academy of Sciences. 2006; 103 (30) 11358–11363.

[40] Okamoto A., Hashimoto K., Nakamura R. Long-range Electron Conduction of *Shewanella* Biofilms Mediated by Outer Membrane C-type Cytochromes. Bioelectrochemistry 2012; 85 61-65.

[41] Bonanni PS., Schrott GD., Robuschi L., Busalmen JP. Charge Accumulation and Electron Transfer Kinetics in *Geobacter sulfurreducens* Biofilms Energy and Environmental Science. 2012; 5 6188–6195.

[42] Nevin KP., Kim BC., Glaven RH., Johnson JP., Woodard TL., Methe BA., DiDonato RJ., Covalla SF., Franks AE., Liu A., Lovley DR. Anode Biofilm Transcriptomics Reveals Outer Surface Components Essential for High Density Current Production in *Geobacter sulfurreducens* Fuel Cells. PLoS ONE 2009; 4(5) 1-10.

[43] Holmes DE., Chaudhuri SK., Nevin KP., Mehta T., Methe BA., Liu A., Ward JE., Woodard TL., Webster J. Lovley DR. Environmental Microbiology. 2006; 8(10) 1805–1815.

[44] Yanga Y., Xua M., Guoa J., Sun G. Bacterial extracellular electron transfer in bioelectrochemical systems. Process Biochemistry. 2012; 47 1707–1714

[45] Okamoto A., Nakamura R., Ishii K., Hashimoto K. In vivo Electrochemistry of C Type Cytochrome-Mediated Electron-Transfer with Chemical Marking. Chem BioChem 2009; 10 2329–2332.

[46] Inoue K., Leang C., Franks AE., Woodard TL., Nevin KP. Lovley DR. Specific Localization of the *c*-Type Cytochrome OmcZ at the Anode Surface in Current-Producing Biofilms of *Geobacter sulfurreducens*. Environmental Microbiology Reports.2011; 3 211–217.

[47] Busalmen JP., Esteve-Nunez A., Berna A., Feliu JM. C-type Cytochromes Wire Electricity-Producing Bacteria to Electrodes. Angewandte Chemie International Edition. 2008; 47 (26) 4874-4877.

[48] Zhao F., Slade RC., Varcoe JR. Techniques for the Study and Development of Microbial Fuel Cells: an Electrochemical Perspective. Chemical Society Review. 2009; 38(7) 1926-1939.

[49] Eaktasang N., Kim D., Lee JW., Park KY., and Kim HS. Enhancement of Electron Transfer by Electrochemical Treatment of Electrode in the Microbial Fuel Cell. In: International Conference on Chemical, Environmental Science and Engineering Pattaya (Thailand).2012; p44-46.

[50] Leang C., Malvankar NS., Franks AE., Nevin KP. Lovley DR. Engineering *Geobacter sulfurreducens* to Produce a Highly Cohesive Conductive Matrix with Enhanced Capacity for Current Production. Energy and Environmental Science. 2013; 6 1901–1908.

[51] Yi H., Nevin KP., Kim B-C., Franks AE., Klimes A., Tender LM., Lovley DR. Selection of a Variant of *Geobacter sulfurreducens* with Enhanced Capacity for Current Production in Microbial Fuel Cells. Biosensors and Bioelectronics. 2009; 24(12) 3498-3503.

[52] El-Naggar MY., Wanger G., Leung KM., Yuzvinsky TD., Southam G., Yang J., Lau WM, Nealson KH., Gorby YA. Electrical Transport Along Bacterial Nanowires from *Shewanella oneidensis* MR-1. Proceedings of the National Academy of Sciences. 2010; 107(42) 18127-18131.

[53] Yuan Y., Zhao B., Zhou S., Zhong S., Zhuang L. Electrocatalytic Activity of Anodic Biofilm Responses to pH Changes in Microbial Fuel Cells. Bioresource Technology 2011; 102 (13) 6887–6891.

[54] Behera MM., Ghangrekar MM.. Performance of Microbial Fuel Cells in Response to Change in Sludge Loading Rate at Different Anodic Feed pH. Bioresource Technology 2009; 100(21) 5114–5121.

[55] Zhuang L., Zhou SG., Li YT., Yuan Y.. Enhanced Performance of Air-Cathode Two-Chamber Microbial Fuel Cells with High-pH Anode and Low-pH Cathode. Bioresource Technology. 2010; 101(10) 3514–3519.

[56] He Z., Huang YL., Manohar AK., Mansfeld F. Effect of Electrolyte pH on the Rate of the Anodic and Cathodic Reactions in an Air-Cathode Microbial Fuel Cell. Bioelectrochemistry 2008; 24 78–82.

[57] Nimje VR., Chen CY., Chen CC., Tsai JY., Chen HR., Huang YM., Jean JS., Chang YF., Shih RC. Microbial Fuel Cell of *Enterobacter cloacae*: Effect of Anodic pH Microenvironment on Current, Power Density, Internal Resistance and Electrochemical Losses. International Journal of Hydrogen Energy. 2011; 36 11093-11101.

[58] Bond DR., Strycharz-Glaven SM., Tender LM., Torres CI. On Electron Transport Tthrough Geobacter Biofilms. ChemSusChem, 2012; 5(6) 1–8.

[59] Mohan SV., Raghavulu SV., Sarma PN. Influence of Anodic Biofilm Growth on Bioelectricity Production in Single Chambered Mediator Less Microbial Fuel Cell Using Mixed Anaerobic Consortia. Biosensors and Bioelectronics 2008; 24(1) 41–47.

[60] Yang S., Du F., Liu H. Characterization of Mixed-Culture Biofilms Established in Microbial Fuel Cells. Biomass and Bioenergy 2012; 46 531-537.

[61] Liu J., Qiao Y., Lu ZS., Song H., Li CM. Enhance Electron Transfer and Performance of Microbial Fuel Cells by Perforating the Cell Membrane. Electrochemistry Communications. 2012; 15(1) 50-53.

[62] Franks AE., Nevin KP., Jia H., Izallalen M., Woodard TL., Lovley DR. Novel Strategy for Three-Dimensional Real-Time Imaging of Microbial Fuel Cell Communities: Monitoring the Inhibitory Effects of Proton Accumulation Within the Anode Biofilm. Energy & Environmental Science. 2009; 2 113-119.

[63] Millo D., Harnisch F., Patil SA., Ly HK., Schröder U., Hildebrandt P. In Situ Spectroelectrochemical Investigation of Electrocatalytic Microbial Biofilms by Surface-Enhanced Resonance Raman Spectroscopy. Angewandte Chemie International Edition. 2011; 50(11) 2625-2627.

[64] Liu J, Qiao Y, Guo CX, Lim S, Song H, Li CM. Graphene/Carbon Cloth Anode for High-Performance Mediatorless Microbial Fuel Cells. Bioresource Technology. 2012; 114 275-280.

[65] Virdis B., Harnisch F., Batstone DJ., Rabaey K., Donose BC. Non-Invasive Characterization of Electrochemically Active Microbial Biofilms Using Confocal Raman Microscopy. Energy and Environmental Science. 2012; 5(5) 7017-7024.

Chapter 5

Removal of COD by Two-Chamber Microbial Fuel Cells

Katalin Belafi-Bako, Balazs Vajda, Peter Bakonyi and
Nandor Nemestothy

Additional information is available at the end of the chapter

http://dx.doi.org/10.5772/58373

1. Introduction

Microbial fuel cells (MFC) are regarded as special bio-electrochemical systems which seem to have a high potential in the future technology for energy production in the section of renewable green energy sources [1-5].

A classical MFC in general [4-8] consist of an anodic and cathodic chamber divided by a proton (cation) selective membrane (Figure 1). Microbes in the anodic cell – mostly attached to the electrode surface forming a biofilm – oxidize the substrates and generate electrons and protons in the process. Then electrons are transported from the anode to the cathode through an external circuit (wire) resulting measurable electric current. Meanwhile protons are passing through the membrane and enter the cathode cell where they combine with oxygen to form water.

MFC-s can be operated by either monoculture (e.g. *Geobacter sulfurreducens* [5], *Shewanella oneidensis, Lactococcus lactis* [9], *Lysinibacillus sphaericus* [10]) – where the electrogenic species release electron to the anode electrode directly or with the use of electroactive metabolites –, or multiculture (microbe consortia) system, which can be found in e.g. anaerobic sewage sludge acting in a similar way. The substrates of the MFCs operated by sewage sludge can be provided from various wastewaters, thus electricity generation in the MFCs can be coupled with the degradation of organic matters even e.g. wastewater treatment processes [11-20].

Many types of wastewaters were investigated (e.g. beer brewery wastewater [22], food industrial wastewaters [22, 23]) where the degradation can be followed by COD removal determination [14]. In MFCs applied for treatment of organic matters, the feed can be enriched with various substrates. It has been noted that non-fermentable substrates are superior to fermentable substrate as the electron donor for power output and electron recovery. E.g. it was reported that an acetate enriched MFC produced higher power output than a glucose enriched

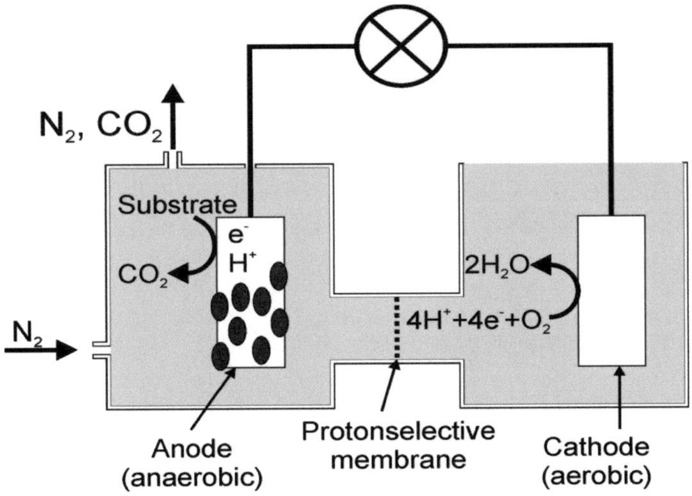

Figure 1. The scheme of a two-chamber microbial fuel cell

MFC [24, 25]. In contrast we found that certain carbohydrate substrates resulted in higher electricity generation [26]. Wastewaters from a sugar beet factory obviously consists of carbohydrates (mono-and/or disaccharides), which are probably suitable enrichments compounds for MFCs.

One of the application segments where extremely promising possibilities are visualised is the integration of the MFCs into wastewater treatment, since these systems are capable to consume organic material even from waste streams to produce power.

In Kaposvar (a city in Southern Hungary) there is a sugar factory, where saccharose (crystallized sugar) is produced from sugar beet [27]. It is an energy intensive process, thus the company has to make great efforts to optimise energy and resource usage, trying to apply environmentally sound, sustainable and waste minimising technologies, which cover water and air quality management as well as usage of by-products. Regarding air quality, the factory has taken extensive measure to decrease dust, smell and sound emissions in recent years, modern biofilters were installed. As to the water issue, special attention is payed to clean the wastewater streams generated during the manufacturing process. The by-products of the processes: the green leaves of sugar beet, the pressed slices (pulp) and other substances are used as components of fertilizers, soil conditioners and for other purposes.

In this factory the processing of all the by-product and waste streams is carefully designed and performed. The leadership of the factory felt responsibility to operate the plant according to environmental safe conditions, therefore they completed the sugar technology with composting – to handle the green parts of the plant –, with a wastewater treatment plant and recently even a biogas fermenter was installed (Figure 2). Inserting microbial fuel cells into this system direct electric power could be produced by reduction of the COD from the waste streams.

http://dx.doi.org/10.5772/58373

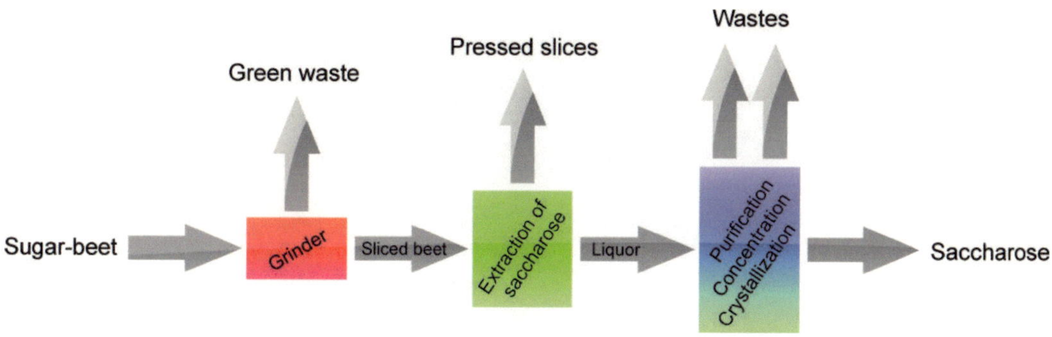

Figure 2. The main process streams in the factory

In this project the aim was to find out how to work and operate a two chamber microbial fuel cell by using wastewater from the sugar beet factory.

2. Materials and methods

A two-chamber MFC was used in the experiments (Figure 3). Both cells have a 240 cm^3 volume, the membrane between them was a Nafion N115 proton selective membrane, its surface area was 50 cm^2 and thickness was 125 μm. In the anodic cell a complex mesophilic anaerobic sludge (from a local working biogas plant, initial COD value was 70 000 mg/dm^3), while in the cathodic cell distilled water was used. Both electrodes were carbon cloth, their surface area was 50 cm^2. The electrodes were connected via a resistor and the voltage data were continuously quantified by a parallel measuring system. It contained a 100 MΩ resistor and the voltage data were directly collected by a data acquisition device (National Instruments USB-6008/6009). The data were recorded by the LabVIEW program. Based on the voltage and resistance data the current values were calculated, thus the electric power (in mW units) was possible to provide. The cumulated electricity data were summarized (in mWh) taken into account the operating time.

The anaerobic sludge applied in the anodic cell was pretreated prior to usage, as it was reported [26]. The MFC was placed in a thermostated container, where the liquids in both cells were possible to circulate and stir. In the cathodic cell air was entered continuously by a pump to ensure aerobic environment, while N_2 was sparged through the anodic cell to assure the anaerobic conditions. The MFC reactor was initially inoculated with the pretreated anaerobic sludge. Then the microbial consortia started to operate in the cell and they were allowed to adapt the actual conditions and colonise locally.

To follow the processes various analytical methods: pH, total solid substance (TSS), COD were applied. COD values of the samples taken from the cells were determined by the potassium dichromate method, which is based on the following oxidation reaction in acidic environment:

$$K_2Cr_2O_7 + 4H_2SO_4 = K_2SO_4 + Cr_2(SO_4)_3 + 4H_2O + 3O$$

Figure 3. The picture of MFC used

The organic substances in the samples were ravaged in the presence of potassium dichromate and catalysts ($AgSO_4$, $HgSO_4$) in sulphuric acid solution at 145 °C. The absorbance of the chromium (III) ions from the reduction of K_2CrO_7 during the oxidation process was determined by a spectrophotometer (Hach LANGE DR 3900) at 600 nm. The COD values of the samples were calculated using a calibration curve taken in the range of 750 – 4000 KOI mg/cm^3.

3. Results

3.1. Experiments with model wastewater

To test the behaviour of the MFC, model substrates were used for the experiments. The composition of the model mixture – similar to the waste stream of the sugar beet factory – was carbohydrates (0.6 %) and inorganic substances (ash, 3 %).

The MFC operation was started by inoculation with the pretreated anaerobic sludge, then the processes were followed by recording the voltage data continuously (Figure 4). A 10 Ω resistor was used in the circuit. After certain intervals 100 cm^3 fresh substrate was added (marked by the arrows) to the system to maintain the carbohydrate concentration. These additions mean 9000 mg/cm^3 COD input each.

The voltage data have shown that the substrate addition enhanced the power generation and quite high voltage (current density) values were achieved (in the range of 10 and 24 mV).

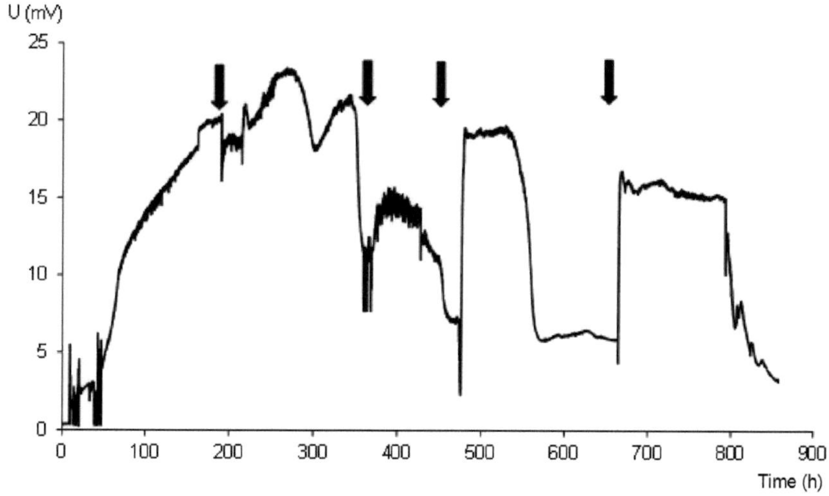

Figure 4. Power generation using model waste water

Samples were taken regularly and the COD values, pH and TSS were determined. The data are summarised in Tables 1 and 2.

Time (h)	COD (mg/dm³)		COD decrease (%)
	Adding substrate	End of period	
1-192	19800	4900	75
192-336	7150	2775	61
336-480	5025	3075	39
480-650	5325	1810	66
650-880	4060	1250	69

Table 1. Changes of COD values using model waste stream

Time (h)	pH	TSS (g/dm³)
1	6.70	35.9
192	8.25	3.9
336	7.65	1.9
480	7.55	1.6
650	7.05	1.1
880	4.45	0.9

Table 2. TSS and pH values using model waste stream

As it can be seen remarkable TSS and COD decrease could be observed during the processes, while electric power was generated continuously. The drop in the pH at 880 h has shown that the operation of the MFC should be ceased due to the unfavourable biochemical processes. The values of current density, J (normalized to the 50 cm^2 carbon cloth electrode) and power density (calculated from the current density and voltage data; Pa and Pm, normalized to the surface of the carbon cloth electrode and the membrane surface area, respectively), moreover the cumulated energy data ∑P (taken into account the time period of the experiment) are summarized in Table 3.

Time (h)	J (mA/m²)	P_a (mW/m²)	P_m (mW/m²)	∑P (mWh/m²)
1-192	260	3.380	4.023	649.0
192-336	416	8.652	10.300	1245.9
336-480	220	2.420	2.880	4114.0
480-650	240	2.880	3.428	4896.0
650-880	300	4.500	5.357	1035.0

Table 3. Results obtained using model waste stream

The values have shown that notable power density was achieved during the operation and finally altogether more than 10 thousands mWh/m^2 energy was generated by consuming the organic substances from the model waste water in the MFC.

3.2. Polarization curve

Our next purpose was to determine the polarization curve using the adapted microbe consortia (grown up in the experiments by the model wastewater). Therefore a serial of experiments were carried out applying various resistances (in the range of 5 and 500 Ω) built in the system and the voltage/current data were measured (Figure 5). Moreover the power density values were calculated.

It can be seen from the figure that voltage data decreased gradually as higher resistances were installed into the system up to 12 mA current, thus the MFC was in a stationer state in the given range from electrochemical point of view. With the growing resistance values, however, increasing power density values were possible to achieve, which mean that it is advisable to use higher resistances for the further measurements. The results shown in the Figure are similar as reported in literature [22].

3.3. Experiments with industrial wastewater

Wastewater from the sugar factory in Kaposvar was used in the next experiments of our work. It consisted of carbohydrates (others than saccharose: mainly glucose and fructose; altogether 0.6 %) and ash 3 %. The other features of the waste water were: pH 5.4, TSS 34.8 g/dm^3 and COD value 12 075 mg/dm^3.

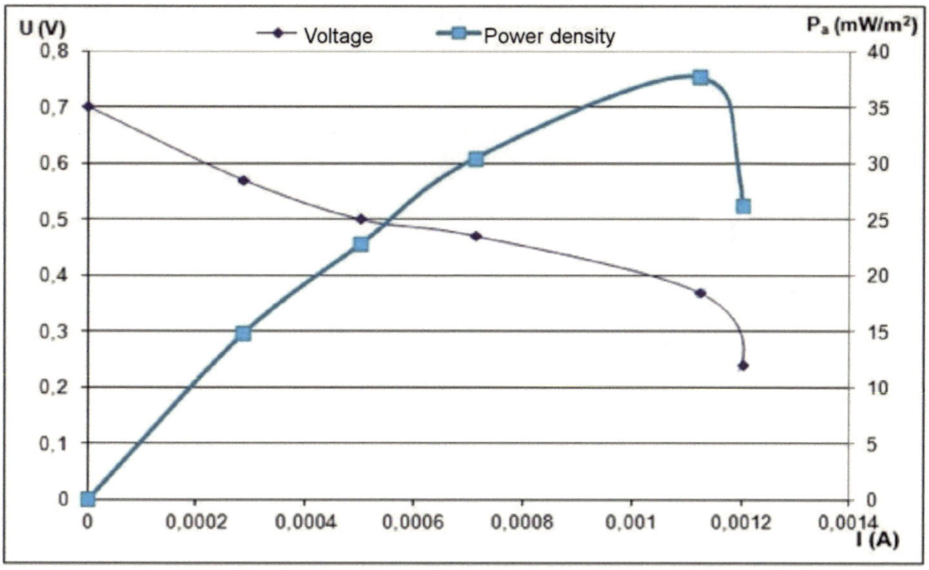

Figure 5. The variance of the voltage and power density in different resistances

Now the MFC was operated by a 330 Ω resistance built in the circuit, which was considered as a more effective operation condition point according to the polarization curve. The adapted microbial consortium was used to inoculate the cell. After certain intervals 100 cm^3 fresh substrate (wastewater) was added (marked by the arrows) to the system. In Figure 6 the voltage data obtained in the MFC by the waste water from the sugar beet factory are presented.

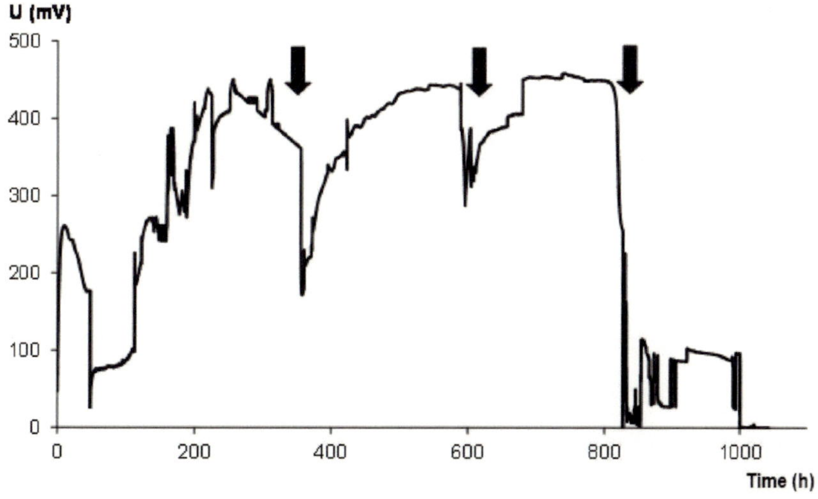

Figure 6. Power generation using industrial waste water from Kaposvar

The changes of COD values of the samples were determined and summarized in Table 4, while the TSS and pH data are shown in Table 5.

Time (h)	COD (mg/dm^3)		decrease of COD (%)
	Adding substrate	End of period	
1-330	21850	13850	37
330-600	13673	5225	62
600-840	7795	3875	50
840-1000	7975	5900	26

Table 4. Changes of COD values using waste water from the factory

Time (h)	pH	TSS (g/dm^3)
1	8.1	34.8
330	8.0	10.8
600	7.6	3.2
840	7.1	2.5
1000	4.1	4.2

Table 5. TSS and pH values using industrial wastewater

The values of current density, power density – which are higher here than in case of model waste water due to the higher resistance – and the cumulated energy data $\sum P$ are listed in Table 6.

Time (h)	J (mA/m^2)	P_a (mW/m^2)	P_m (mW/m^2)	$\sum P$ (mWh/m^2)
1-330	181	52	59.5	17 160
330-600	242	95	113	25 650
600-840	245	99.2	118.2	23 808
840-1000	51	4.2	5.0	642

Table 6. Results obtained using industrial waste stream

The current and power density values were outstanding compared to literature data [20-22] or the values obtained with the model wastewater. Altogether more than 60 thousands mWh/m^2 energy was generated during the over 800 hour operation. In the last interval (after 840 hour) the substrate addition did not result in any increase of voltage data, moreover the pH

http://dx.doi.org/10.5772/58373

decreased sharply, showing that the biochemical processes in the MFC have not been working properly any longer, thus the operation was ended.

4. Conclusion

A two-chamber MFC was designed and built in our laboratory where carbon cloth (fibres) were used as electrodes and Nafion 0125 protonselective membrane was placed between the two cells. Pretreated anaerobic sludge from a local biogas plant was applied in the anode chamber. Experiments with model and industrial wastewaters from a sugar beet factory were carried out to study the efficiency and stability of the system and to measure the electric power generated. It has turned out that it was possible to generate bioelectricity from the particular wastewater and the power data obtained were found similar or higher to the data published in literature, i.e. comparable performance was achieved in our MFC system. Our next purpose is to study the MFC for longer term operation, even in continuous mode of operation, so that we can insert the MFC into the processing scheme of the sugar beet factory [28].

Acknowledgements

This work was supported by the European Union and co-financed by the European Social Fund in the frame of the TAMOP-4.2.2.A-11/1/KONV-2012-0071 and TÁMOP-4.2.2.A-11/1/KONV-2012-0038 projects; and the János Bolyai Research Scholarship of the Hungarian Academy of Sciences.

Author details

Katalin Belafi-Bako*, Balazs Vajda, Peter Bakonyi and Nandor Nemestothy

*Address all correspondence to: bako@almos.uni-pannon.hu

University of Pannonia, Veszprem, Hungary

References

[1] B.E. Logan: Microbial fuel cells, Wiley & Sons, New York, 2008

[2] www.microbialfuelcell.org

[3] Y. Mohan, S.N. Muthu Kumar, D. Das: Electricity generation using microbial fuel cells, Int. J. Hydrogen En. 33 (2008) 423-426

[4] K. Rabaey, W. Verstraete: Microbial fuel cells: novel biotechnology for energy generation, Trends in Biotechnology, 23 (2005) 291-298

[5] D.R. Bond, D.R. Lovley: Electricity production by *Geobacter sulfurreducens* attached to electrodes, Appl. Env. Microbiol. 69 (2003) 1548-1555

[6] B.E. Logan, J.M. Regan: Electricity-producing bacterial communities in microbial fuel cells, Trends in Microbiology, 14 (2006) 512-518

[7] B.E. Logan: Microbial fuel cells – Challenges and Applications, Environmental Science and Technology 40 (2006) 5161-5586

[8] M. Zhou, T. Jin, Z. Wu, M. Chi, T. Gu: Mirobial fuel cells for bioenergy and bioproducts, Green Energy and Technology 62 (2012) 131-171

[9] M.A. Rosenbaum, H.Y. Bar, Q.K. Beg, D. Segré, J. Booth, M.A. Cotta, L.T. Angenent: *Shewanella oneidensis* in a lactate-fed pure-culture and a glucose-fed co-culture with *Lactococcus lactis* with an electrode as electron acceptor, Bioresource Technology 102 (2011) 2623-2628

[10] A. Nandy, V. Kumar, P. P. Kundu: Utilization of proteinaceous materials for power generation in a meidatorless microbial fuel cell b a new electrogenic bacteria *Lysinibacillus sphaericus* VA5, Enzyme and Microbial Technology 53 (2013) 339-344

[11] A.K. Mungray, Z.V.P. Murthy, A.J. Tirpude: Post treatment of up-flow anaerobic sludge blanket based sewage treatment plant effluents: A review, Desalination and Water Treatment 22 (2010) 220-237

[12] H. Moon, I.S. Chang, B.H Kim: Continuous electricity production from artificial wastewater using a mediator-less microbial fuel cell, Bioresource Technology 97 (2006) 621-627

[13] Z. Du, H. Li, T. Gu: A state of the art review on microbial fuel cells: A promising technology for wastewater treatment and bioenergy, Biotechnology Advances 25 (2010) 464-482

[14] H. Moon, S.I. Chang, K.K. Jang, B.H. Kim: Residence time distribution in microbial fuel cell and its influence on COD removal with electricity generation, Biochemical Engineering Journal 27 (2005) 59-65

[15] D.H. Park, J.G. Zeikus: Improved fuel cell and electrode designs for producing electricity from microbial degradation, Biotechnol. Bioeng. 81 (2003) 348-355

[16] D.K. Daniel, B.D. Mankidy, K. Ambarish, R. Manogary: Construction and operation of a microbial fuel cell for electricity generation from wastewater, Int. J. Hydrogen En. 34 (2009) 7555-7560

[17] I. Ivanov, L. Ren, M. Siegert, B.E. Logan: A quantitative method to evaluate microbial electrolysis cell effectiveness for energy revovery and wastewater treatment, In. J. Hydrogen Energy 38 (2013) 13135-13142

[18] R.A. Rozendal, H.V,M. Hamalers, K. Rabaey, J. Keller, C.J.N. Buisman: Towards practical implementation of bioelectrochemical wastewater treatment, Trends in Biotechnology 26 (2008) 450-459

[19] G. Velvizhi, R. K. Goud, S.V. Mohan: Anoxic bio-electrochemical system for treatment of complex chemical wastewater with simultaneous bioelectricity generation, Bioresource Technology, 151 (2014) 214-220

[20] Y. Tian, C. Ji, K. Wang, P. Le-Clech: Assessment of an anaerobic membrane bio-electrochemical reactor for wastewater treatment and energy recovery, J. Membrane Sci., 450 (2014) 242-248

[21] Q. Wen, Y. Wu, D. Cao, L. Zhao, Q. Sun: Electricity generation and modelling of microbial fuel cell from continuous beer brewery wastewater, Bioresource Technology 100 (2009) 4171-4175

[22] B Cercado-Quezada, M. L. Delia, A. Bergel: Testing various food-industry wastes for electricity production in microbial fuel cell, Bioresource Technology 101 (2010) 2748-2754

[23] X. M. Li, K. Y. Cheng, J.W.C. Wong: Bioelectricity production from food waste leachate using microbial fuel cells: Effect of NaCl and pH, Bioresource Technology 149 (2013) 452-458

[24] D. Pant, G. van Bogaert, L. Diels, K. Vanbroekhoven: A review of the substrates used in microbial fuel cells (MFCs) for sustainable energy production, Bioresource Technology 101 (2010) 1533-1543

[25] Y. Yuan, S. Zhou, N. Xu, L. Zhuang: Electrochemical characterization of anodic biofilm enriched with glucose and acetate in single-chamber microbial fuel cells, Colloid and Surfaces B: Biointerfaces 82 (2011) 641-646

[26] Bélafi-Bakó, K., Vajda, B., Nemestóthy, N.: Study on operation of a microbial fuel cell using mesophilic anaerobic sludge, *Desalination and Water Treatment*, 35 (2011) 222-226

[27] www.agrana.hu

[28] D. Akman, K. Cirik, S. Ozdemir, B. Ozkaya, O. Cinar: Bioelectricity generation in continuously-fed microbial fuel cell: Effects of anode electrode material and hydraulic retention time, Bioresource Technology 149 (2013) 459-464